Energy Science, Engineering and Technology

Electrical Engineering Developments

www.novapublishers.com

Energy Science, Engineering and Technology

Integrated Energy Systems: Design, Control and Operation
Bikram Das (Editor)
Abanishwar Chakraborti (Editor)
Arvind Kumar Jain (Editor)
Subhadeep Bhattacharjee (Editor)
2023. ISBN: 979-8-89113-206-1 (Softcover)
2023. ISBN: 979-8-89113-227-6 (eBook)

Photovoltaic Systems: Advances in Research and Applications
Sudip Mandal, PhD (Editor)
Pijush Dutta, PhD (Editor)
2023. ISBN: 979-8-89113-102-6 (eBook)

More information about this series can be found at
https://novapublishers.com/product-category/series/energy-science-engineering-and-technology/

Electrical Engineering Developments

Design and Simulation of Electrical Machines with MATLAB
L. Ashok Kumar, PhD (Author)
S. Albert Alexander, PhD (Author)
Y. Uma Maheswari, PhD (Author)
2021. ISBN: 978-1-68507-411-1 (Hardcover)
2021. ISBN: 978-1-68507-476-0 (eBook)

Planar Antenna: Design, Fabrication, Testing, and Application
Praveen Kumar Malik, PhD (Editor)
2021. ISBN: 978-1-53619-898-0 (Hardcover)
2021. ISBN: 978-1-68507-042-7 (eBook)

More information about this series can be found at
https://novapublishers.com/product-category/series/electrical-engineering-developments/

Suresh Nagappan Sundaram
N. S. Padmavathy
Mohit Hemanth Kumar
A. Joseph Godfrey
Editors

Electric Vehicle Technology Structure, Instrumentation and Challenges

https://doi.org/10.52305/RVBV2807

NOTICE TO THE READER

Library of Congress Cataloging-in-Publication Data

ISBN: 979-8-89113-695-3 (Softcover)
ISBN: 979-8-89113-784-4 (eBook)

Published by Nova Science Publishers, Inc. † New York

Contents

Preface

The automotive industry is undergoing a remarkable transformation, driven by the urgent need to reduce greenhouse gas emissions, combat air pollution, and transition toward more sustainable transportation methods. Hybrid Electric Vehicles (HEVs) represent a paradigm shift in this landscape, leveraging two or more power sources to enhance efficiency, curtail emissions, and optimize energy consumption. Hybrid Electric Vehicles (HEVs) are a significant advancement in the automotive industry, offering a practical and eco-friendly solution to the challenges of fuel efficiency and emissions reduction. They also play a pivotal role in the transition to sustainable transportation. HEVs are ideally suited for city driving, where frequent stops and slow-moving traffic are routine. When operating in electric mode, HEVs generate minimal noise and zero tailpipe emissions, contributing to a quieter and cleaner urban environment. As technology continues to advance, HEVs are likely to play an increasingly significant role in modern transportation systems. This comprehensive exploration of HEVs will delve into their definition, various types, significance in modern transportation, and adaptation to evolving regulatory frameworks. We will also examine the challenges and opportunities associated with HEVs, offering a holistic perspective on this revolutionary technology.As we embark on this journey, it is essential to acknowledge the pioneering efforts of scientists, engineers, and policymakers who have paved the way for HEVs. Their dedication and vision have laid the foundation for a more sustainable future of transportation. We invite you to join us in exploring this fascinating topic and discover the potential of HEVs to transform the way we travel.

Chapter 1

Paving the Way for Electric Vehicle Infrastructure Worldwide

Suresh Nagappan Sundaram*
Department of Electrical and Electronics Engineering,
Saveetha School of Engineering,
Saveetha Institute of Medical and Technical Sciences, Saveetha University,
Chennai, Tamil Nadu, India

Introduction

The global landscape of transportation is currently undergoing a significant transformation with the rapid growth of electric vehicles (EVs). This shift holds the promise of mitigating climate change, reducing greenhouse gas emissions, and enhancing energy security. The development of a comprehensive and robust infrastructure for electric vehicles worldwide is essential to realize these benefits. In this essay, we will delve into the multifaceted aspects of this infrastructure, which includes charging networks, battery technology, policy initiatives, and their collective impact on the widespread adoption of electric vehicles.

Charging Infrastructure

At the heart of the EV infrastructure is the charging network. The widespread adoption of electric vehicles depends on the availability of efficient and accessible charging solutions. The charging infrastructure encompasses three primary types of charging stations:

---

* Corresponding Author's Email: drsureshns@gmail.com.

In: Electric Vehicle Technology Structure, Instrumentation and Challenges
Editors: S. N. Sundaram, P. N. Sundaram, M. H. Kumar et al.
ISBN: 979-8-89113-695-3

- *Residential Charging:* The simplest form of EV charging occurs at home. Residential charging relies on standard electrical outlets (Level 1) or more potent Level 2 chargers, enabling EV owners to conveniently charge their vehicles overnight for everyday use.
- *Public Charging:* Public charging stations are critical for longer journeys and urban areas where residents may lack access to private charging facilities. Level 2 charging stations and Level 3 DC fast chargers are strategically placed at various locations, including parking lots, shopping centers, and highways.
- *Ultra-Fast Charging*: Ultra-fast chargers, exemplified by Tesla's Superchargers and the Electrify America network, offer high-speed charging options, significantly reducing charging times. These are indispensable for long-distance travel and enhancing the appeal of EVs.

Charging infrastructure should aim to be universally accessible, cost-effective, and swift, facilitating the seamless integration of electric vehicles into daily life. Governments, automakers, and private companies have invested significantly in expanding and improving these networks.

Battery Technology

The progression of battery technology is pivotal to the success of EVs. Key aspects of battery development include energy density, charging speed, cost efficiency, and durability. While lithium-ion batteries have dominated the market, ongoing research and innovation are focused on areas such as solid-state batteries, which hold the potential to revolutionize the EV industry.

- *Energy Density*: Elevating the energy density of batteries allows for extended driving ranges on a single charge. Advances in materials, such as high-nickel cathodes and silicon anodes, have played a crucial role in achieving this objective.
- *Charging Speed*: Speedier charging times are imperative for the convenience of electric vehicles. Research into ultra-fast charging and effective battery thermal management systems is continuously advancing.

- *Cost Reduction:* The reduction in battery production costs is pivotal to making EVs more affordable for consumers. This is achieved through economies of scale, recycling, and sustainable mining practices.
- *Durability:* Batteries must endure thousands of charging cycles to ensure the longevity of EVs. Progress in materials and manufacturing techniques has enhanced battery life and reliability.

Policy Initiatives

Government policies wield significant influence in shaping the EV infrastructure. Numerous countries and regions have implemented a range of incentives and regulations to encourage the adoption of electric vehicles.

- *Purchase Incentives:* Tax credits, rebates, and grants are offered to consumers to alleviate the initial cost of purchasing an EV, making them more attractive and affordable.
- *Charging Infrastructure Support:* Governments often invest in building charging infrastructure and provide financial incentives to businesses to install charging stations.
- *Emission Regulations*: Stricter emissions standards incentivize automakers to produce more electric vehicles and reduce the production of traditional internal combustion engine (ICE) vehicles.
- *Clean Energy Transition:* Many countries are transitioning their power grids to rely more on renewable energy sources, ensuring that EVs have a smaller carbon footprint.

Global Impact

The development of a comprehensive electric vehicle infrastructure extends its impact on a global scale:

- *Environmental Benefits*: The widespread adoption of EVs holds the potential to significantly reduce greenhouse gas emissions and improve air quality, contributing to global efforts to combat climate change.

- *Energy Security:* Decreasing reliance on fossil fuels enhances energy security by reducing vulnerability to oil price fluctuations and geopolitical conflicts.
- *Economic Growth*: The EV industry has the potential to create jobs and stimulate economic growth in manufacturing, research, and development.
- *Technological Advancement:* The EV sector drives innovation in battery technology, energy storage, and grid management, spurring technological advancements in related fields.

Conclusion

The infrastructure for electric vehicles is a linchpin in the transition toward a more sustainable and environmentally friendly future. A well-developed charging network, advancements in battery technology, and supportive policies are all critical components. The global effort to build this infrastructure not only accelerates the adoption of EVs but also fosters economic growth and environmental stewardship. As the world increasingly recognizes the importance of electric vehicles, we move closer to realizing a more sustainable and clean transportation future.

Advancing Electric Vehicle Infrastructure in the American Countries

Introduction

The global shift toward electric vehicles (EVs) as a sustainable and eco-friendly mode of transportation is gaining momentum, and countries across the Americas are actively contributing to this transformative journey. In this essay, we will explore the pivotal role played by American nations in establishing a robust and comprehensive EV infrastructure. This infrastructure is vital for supporting the growth of electric vehicles and reducing the environmental impact of transportation.

Charging Infrastructure

At the core of the EV revolution in the Americas lies the development of a comprehensive charging infrastructure. Key components of this infrastructure include:

1. *Urban Charging Stations*: American cities, known for their diversity and density, require strategically placed urban charging stations. These stations must be accessible in residential areas, commercial districts, and public parking facilities to cater to urban residents' needs.
2. *Interurban Fast Charging Networks*: To facilitate long-distance travel, American countries have invested in interurban fast charging networks. These networks line major highways and critical transportation routes, alleviating range anxiety and making electric vehicles a practical choice for extensive journeys.
3. *Home Charging Solutions*: Many Americans have the advantage of private garages or parking spaces. Therefore, there is a push for residential charging solutions, often supported by government incentives for the installation of home charging stations.
4. *Cross-Border Compatibility:* American nations are diligently working on standardizing charging connectors and protocols, ensuring that electric vehicles can easily traverse borders and utilize charging networks across different countries. This standardization promotes electric vehicle adoption at a continental level.

Government Initiatives

Government policies and initiatives are pivotal in shaping the EV infrastructure. American countries have adopted a range of measures to foster electric vehicle growth:

1. *Incentives and Subsidies:* To incentivize consumers to purchase electric vehicles, governments offer financial benefits such as tax breaks, rebates, and subsidies, enhancing the affordability of EVs.
2. *Charging Infrastructure Funding*: Public investments in charging infrastructure are a common practice in the Americas. Governments frequently collaborate with private enterprises to expand the network and alleviate the financial burden on businesses.

3. *Emission Reduction Goals*: Numerous American nations have set ambitious goals for reducing carbon emissions. These objectives drive the transition to electric vehicles and spur the development of associated infrastructure.
4. *Regulatory Framework*: Regulations that support EV adoption, including stringent emissions standards and fuel economy regulations, create a conducive environment for automakers to manufacture and promote electric vehicles.

Battery Technology and Manufacturing

The Americas are making substantial strides in battery technology and manufacturing:

1. *R&D Investments*: American countries and regions allocate resources to research and development in battery technology, focusing on enhancing energy density, charging speed, and cost efficiency.
2. *Local Battery Production*: Several American countries invest in domestic battery manufacturing to secure the supply chain and reduce reliance on foreign imports.
3. *Sustainability and Recycling*: Sustainability is a core focus in the Americas, with recycling programs emerging to ensure the eco-friendliness of electric vehicles throughout their lifecycle.

Unique Challenges

Despite the numerous advantages, American countries face distinct challenges in the transition to electric vehicles. These include concerns related to electricity grid capacity, charging infrastructure in rural and remote areas, and the impact on employment in the traditional automotive industry.

Conclusion

The Americas are actively leading the global shift toward electric vehicles. Establishing a robust EV infrastructure is not only essential for reducing carbon emissions and addressing climate change but also vital for fostering the growth of a sustainable, green transportation sector and the creation of cleaner and more sustainable economies throughout the continent. Through strategic investments in charging infrastructure, government support, and advancements in battery technology and local manufacturing, American countries are paving the way for a future in which electric vehicles are a common sight on the roads, revolutionizing the transportation sector and contributing to a cleaner and more sustainable future.

Advancing Electric Vehicle Infrastructure in European Nations

Introduction

The global shift towards electric vehicles (EVs) as an eco-friendly and sustainable transportation option is gaining momentum, and European countries are taking the lead in this transformative journey. This essay delves into the pivotal role played by European nations in establishing an effective and widespread EV infrastructure. This infrastructure is the linchpin for supporting the growth of electric vehicles and reducing the environmental footprint of transportation.

Charging Infrastructure

At the core of the EV revolution in European nations lays the development of a comprehensive charging infrastructure. Essential elements of this infrastructure include:

1. *Urban Charging Stations:* European cities, often characterized by their high population density, require strategically positioned urban charging stations. These stations must be accessible in residential areas, commercial hubs, and public parking facilities to cater to urban residents' needs.
2. *Interurban Fast Charging Networks*: To facilitate long-distance travel, European countries have made substantial investments in interurban fast charging networks. These networks line major highways and crucial transportation routes, alleviating range anxiety and making electric vehicles a practical choice for extensive journeys.

3. *Home Charging Solutions*: Many European residents have the advantage of private garages or parking spaces. Therefore, there is a push for residential charging solutions, often supported by government incentives for the installation of home charging stations.
4. *Cross-Border Compatibility*: European nations are diligently working on standardizing charging connectors and protocols, ensuring that electric vehicles can easily traverse borders and utilize charging networks across different countries. This harmonization is pivotal in promoting electric vehicle adoption at a continental level.

Government Initiatives

European governments wield significant influence in shaping the EV infrastructure. They have adopted a range of policies and initiatives to foster electric vehicle growth:

1. *Incentives and Subsidies*: To incentivize consumers to purchase electric vehicles, governments offer financial benefits such as tax breaks, rebates, and subsidies, thereby enhancing the affordability of EVs.
2. *Charging Infrastructure Funding*: Public investments in charging infrastructure are a common practice in Europe. Governments frequently collaborate with private enterprises to expand the network and alleviate the financial burden on businesses.
3. *Emission Reduction Goals*: Numerous European nations have set ambitious goals for reducing carbon emissions. These objectives propel the transition to electric vehicles and spur the development of associated infrastructure.
4. *Regulatory Framework*: Regulations that support EV adoption, including stringent emissions standards and fuel economy regulations, create a conducive environment for automakers to manufacture and promote electric vehicles.

Battery Technology and Manufacturing

Europe stands as a hub for battery technology and manufacturing, making significant strides in:

1. *R&D Investments*: European countries and the European Union allocate resources to research and development in battery technology, focusing on enhancing energy density, charging speed, and cost efficiency.
2. *Local Battery Production*: Several European countries, including Germany, Sweden, and France, invest in domestic battery manufacturing to secure the supply chain and reduce reliance on foreign imports.
3. *Sustainability and Recycling*: Europe is steadfast in its commitment to sustainable battery production and recycling. Recycling programs are emerging to ensure the eco-friendliness of electric vehicles throughout their lifecycle.

Unique Challenges

Despite the numerous advantages, European nations face distinct challenges in the transition to electric vehicles. These encompass concerns regarding electricity grid capacity, charging infrastructure in rural areas, and the impact on employment in the traditional automotive industry.

Conclusion

European nations lead the way in the global shift towards electric vehicles. The establishment of a robust EV infrastructure is not only essential for reducing carbon emissions and addressing climate change but also critical for the rejuvenation of the European automotive industry and the promotion of sustainable and eco-friendly transportation. Through strategic investments in charging infrastructure, government support, and advancements in battery technology and local manufacturing, European countries are paving the way for a future in which electric vehicles are ubiquitous. This transition will not only revolutionize the transportation sector but also contribute to cleaner and more sustainable economies across the continent.

Accelerating Electric Vehicle Infrastructure Development in Asian Countries

Introduction

The global transition to electric vehicles (EVs) is gaining ground as the world seeks sustainable and eco-friendly transportation solutions. Asian countries, characterized by their vast populations and rapidly growing economies, play a central role in this transformation. The development of an efficient and widespread EV infrastructure in Asia is pivotal to realizing the potential of electric vehicles. This essay will explore the unique challenges and opportunities facing Asian nations as they work towards creating a robust infrastructure to support the growth of electric vehicles.

Charging Infrastructure

Charging infrastructure stands as the bedrock of the EV ecosystem. In Asian countries, the development of charging networks is shaped by specific factors, including population density, urbanization, and existing energy structures. To drive the adoption of EVs, Asian countries must focus on several key aspects of charging infrastructure:

1. *Urban Charging Hubs*: In densely populated Asian cities, urban charging hubs are indispensable. These hubs must be strategically placed in residential areas, business districts, and commercial centers to cater to urban residents' needs.
2. *Interurban Charging Networks*: The establishment of efficient charging networks connecting cities and regions is crucial for long-distance travel. Investments in fast-charging stations along highways and intercity routes are vital to alleviate range anxiety.
3. *Apartment and Condo Charging*: In countries like Japan and South Korea, characterized by high-rise apartment living, the installation of charging facilities in residential complexes is a necessity. This ensures that EV ownership remains a practical choice for urban populations.
4. *Shared Infrastructure:* Collaboration among Asian countries can result in shared charging infrastructure standards, making cross-

border EV travel and trade more seamless. This approach promotes electric vehicle adoption on a regional scale.

Government Initiatives

Government policies and initiatives are key drivers in the development of EV infrastructure. Asian countries have implemented diverse approaches to support electric vehicle growth:

1. *Incentives and Subsidies*: Financial incentives such as tax breaks, rebates, and subsidies are offered by many Asian governments to stimulate EV purchases, making electric vehicles more affordable for consumers.
2. *Charging Infrastructure Investment*: Governments are investing in the expansion of charging networks, both for public and private sectors. Such investments ease the financial burden on private companies and promote the development of charging stations.
3. *Emission Reduction Targets*: Several Asian nations have set ambitious targets to reduce carbon emissions. These goals serve as strong motivators for the transition to electric vehicles and drive the development of associated infrastructure.
4. *Regulatory Support*: Regulations supporting EV adoption, including fuel economy standards and emissions limits, create favourable conditions for automakers to manufacture and market electric vehicles.

Battery Technology and Local Manufacturing

While battery technology is not unique to Asia, the region plays a significant role in battery production and innovation. Countries like China and South Korea are leaders in battery manufacturing. To accelerate electric vehicle infrastructure development, Asian countries must consider the following:

1. *Investing in R&D*: Continued investments in research and development are essential to improving battery technology, increasing energy density, and reducing costs.

2. *Local Battery Production*: Encouraging domestic battery production enhances supply chain security and reduces dependence on foreign imports. China, for example, has made substantial investments in building a strong domestic battery industry.
3. *Recycling Initiatives*: Developing recycling and disposal programs for batteries is vital to ensure that electric vehicles remain environmentally friendly throughout their lifecycle.

Unique Challenges

Asia faces unique challenges in the quest to develop electric vehicle infrastructure. These challenges include managing rapid urbanization, addressing air pollution concerns, and ensuring that charging networks are tailored to the diverse transportation needs of the continent.

Conclusion

Asian countries are leading the global charge in the transition to electric vehicles. The development of a robust EV infrastructure in this region is not only necessary for reducing carbon emissions and combating climate change but also essential for addressing air pollution and enhancing energy security. With strategic investments in charging infrastructure, government support, and advancements in battery technology and local manufacturing, Asian countries can accelerate the shift towards a future where electric vehicles are the norm. This transition will not only reshape the transportation sector but also contribute to sustainable and greener economies throughout the continent.

Empowering the Future: Electric Vehicle Infrastructure in Australia

Introduction

Amid a global transition towards sustainable transportation, electric vehicles (EVs) have emerged as a promising solution to combat climate change and reduce carbon emissions. In Australia, a nation characterized by vast landscapes and unique geographical challenges, the development of a comprehensive and robust EV infrastructure is pivotal to realizing a sustainable future of transportation. This essay delves into the evolution, challenges, and opportunities surrounding electric vehicle infrastructure in Australia.

Charging Infrastructure

The cornerstone of EV adoption in Australia is the establishment of a well-rounded charging infrastructure. Key elements of this infrastructure include:

1. *Urban Charging Stations:* Australian cities require strategically located urban charging stations to cater to the daily charging needs of urban residents and those without home charging capabilities.
2. *Interurban Fast Charging Networks*: The expansive distances between cities necessitate the creation of interurban fast charging networks, allowing for long-distance travel and alleviating range anxiety.
3. *Home Charging Solutions*: Given that many Australians have access to private garages and parking spaces, the availability of home charging solutions is crucial. Government incentives further encourage the installation of residential charging infrastructure.
4. *Remote Area Charging:* Australia's vast and remote regions require charging solutions that cater to long-distance travellers and residents in areas with limited infrastructure.

Government Initiatives

Government policies and incentives play a significant role in shaping the EV infrastructure in Australia:

1. *Incentives and Rebates*: Financial incentives, tax breaks, and rebates are offered to stimulate EV adoption, rendering electric vehicles more financially accessible for Australian consumers.
2. *Charging Infrastructure Investment*: Governments are actively investing in expanding the charging network and partnering with private businesses to establish more charging stations. These efforts help alleviate the financial burden on the private sector.

3. *Emission Reduction Goals*: Australia has set emissions reduction targets, facilitating the transition to electric vehicles and encouraging the development of infrastructure to support clean transportation.
4. *Regulatory Framework*: Regulations, including more stringent emissions standards and fuel economy requirements, create a conducive environment for automakers to manufacture and promote EVs.

Battery Technology and Manufacturing

Australia is making substantial progress in battery technology and local manufacturing:

1. *Research and Development:* The country is directing investments toward research and development to enhance battery technology, with a focus on increasing energy density, charging speed, and cost efficiency.
2. *Local Battery Production*: Efforts are underway to establish local battery manufacturing facilities to secure the supply chain and reduce dependence on battery imports.
3. *Sustainability and Recycling*: Australia is committed to sustainable battery production and recycling, ensuring the environmental friendliness of electric vehicles throughout their lifecycle.

Unique Challenges

Australia's unique challenges in developing EV infrastructure encompass addressing vast distances between cities, ensuring charging infrastructure in remote areas, and transitioning to a cleaner energy grid, given that much of the nation still relies on fossil fuels for electricity generation.

Conclusion

Australia is making significant strides in the development of a robust electric vehicle infrastructure. The establishment of such infrastructure is pivotal for reducing carbon emissions, mitigating climate change, and reshaping the transportation sector into one that is cleaner and more sustainable. Through strategic investments in charging infrastructure, government support, and advancements in battery technology and local manufacturing, Australia is paving the way for a future where electric vehicles play a significant role in reducing the environmental impact of transportation.

Driving Forward: The Electric Vehicle Infrastructure in China

Introduction

China has emerged as a global leader in the electric vehicle (EV) revolution, committing to reshape its transportation landscape and combat environmental challenges head-on. The development of a comprehensive EV infrastructure serves as a foundation for this transformative journey. In this essay, we explore the evolution, challenges, and opportunities surrounding the electric vehicle infrastructure in China.

Charging Infrastructure

The crux of EV adoption in China lies in the continuous expansion of charging infrastructure. Key elements of this infrastructure encompass:

1. *Urban Charging Networks*: Chinese cities boast a plethora of strategically positioned urban charging stations, serving residential areas, commercial hubs, and public parking facilities to meet the charging needs of urban residents.
2. *High-Speed Charging Corridors*: China has made substantial investments in high-speed charging networks along highways and intercity routes, alleviating range anxiety and facilitating long-distance EV travel.
3. *Residential Charging*: Home charging solutions are prevalent, with a significant proportion of urban residents having access to private parking and charging amenities.

4. *Shared Charging Facilities*: Collaborative endeavors between public and private entities have given rise to shared charging infrastructure, fostering standardization and cross-network utilization.

Government Initiatives

Government policies and incentives are the driving force behind the rapid expansion of the EV infrastructure in China:

1. *Incentives and Subsidies*: Financial incentives, tax incentives, and rebates are deployed to stimulate EV adoption, rendering electric vehicles more cost-effective for consumers.
2. *Charging Infrastructure Investment*: The government allocates significant funds to support the expansion of the charging network, working closely with private enterprises to share the financial burden and spur infrastructure growth.
3. *Emission Reduction Goals*: China has set ambitious emissions reduction targets, propelling the transition to electric vehicles and nurturing the development of supporting infrastructure.
4. *Regulatory Support*: Stringent emissions standards and fuel economy regulations provide a conducive environment for automakers to manufacture and promote EVs.

Battery Technology and Manufacturing

China plays a central role in battery technology and local manufacturing:

1. *Research and Development*: Substantial investments are channeled into research and development in battery technology, focusing on improving energy density, charging speed, and cost efficiency.
2. *Local Battery Production*: China has cultivated a thriving domestic battery industry, ensuring a secure supply chain and diminishing reliance on foreign imports.
3. *Sustainability and Recycling*: China is committed to sustainable battery production and recycling, guaranteeing that electric vehicles remain environmentally friendly throughout their lifecycle.

Unique Challenges

China grapples with unique challenges in its quest to develop a comprehensive EV infrastructure, including managing rapid urbanization, addressing air pollution concerns, and ensuring that charging networks are equipped to handle the diverse transportation needs of the nation.

Conclusion

China is steering the global transition towards electric vehicles. The establishment of a robust EV infrastructure is not only essential for reducing carbon emissions and addressing environmental concerns but also for cultivating a sustainable and innovative transportation sector. Through strategic investments in charging infrastructure, government support, and advancements in battery technology and local manufacturing, China is at the forefront of a future where electric vehicles are ubiquitous, reshaping the transportation sector and contributing to cleaner and more sustainable urban environments.

Analysis of Roof Top Solar PV System Available for EV - Grid Integration in India

Introduction

India's power sector has a total installed capacity of approximately 1,46,753 Megawatt (MW) of which 54% is coal-based, 25% hydro, 8% is renewable's and the balance is the gas and nuclear-based. Power shortages are estimated at about 11% of total energy and 15% of peak capacity requirements which islikely to increase in the coming years. India's current economic growth rate of 7.4% makes it, along with China, the fastest growing major economy in the world. India now is a two trillion dollar economy, taking less than ten years to double in size [1]. In another milestone, the National Population Stabilization Fund reported that the population of India had crossed the 1.27 billion mark and was on track to become the most populous nation on earth by 2050, surpassing China [2].

India is on a growth trajectory, and its developmental aspirations demand a commensurate ability on the part of the energy sector to meet its developmental goals without losing sight of environmental sustainability goals. This double requirement has put the focus on renewable energy sources like wind and solar.

One analyst recently observed that though the country is rich in coal, it is also abundantly capable with renewable energy in the form of solar, wind, hydro and bioenergy. Future economic growth crucially depends on the long-term availability of energy from sources that are affordable, accessible and environmentally friendly [3, 4].

Mindful of the urgency and the opportunity, since at least 2008, there has been a big push in the Indian policy circles to "go green," that is, to ramp up the contribution of renewable energy sources to the domestic power supply. India's plans for renewable generation are ambitious and aggressive. By the end of the 11th Five Year Plan, the stated target is addition of 2,000 MW of capacity from renewable sources. The target for the 12th Five Year Plan is 36GW, where in the quantum of energy from renewable sources is expected to increase and the share of renewables (including hydro) to 20% from the current 12% [5].

The Indian solar energy sector has been growing rapidly, in the past few years, majorly due to Government's initiatives such as tax exemptions and subsidies. Due to technical potential of 5,000 trillion kWh per year and minimum operating cost, Solar Power is considered the best suited energy source for India.

On the demand side, adoption of several smart grid pilots and rooftop solar power generation is being encouraged by central and state authorities through subsidies and other incentives. One major impetus for these "go green" measures came in the shape of the National Action Plan for Climate Change (NAPCC). Released in 2008, the Plan articulated a national strategy for achieving a high economic growth rate in an environmentally friendly manner as India's pre-emptive response to the deleterious impacts of greenhouse gas (GHG) emissions on climate change. Accordingly, the Government of India's Jawaharlal Nehru National Solar Mission (JNNSM, also called "National Solar Mission"), one of eight "missions," was inaugurated as part of the Action Plan with the target of 20 Giga watts of grid-connected solar power generation capacity by the year 2022 (Table 1).

Table 1. Calculation of solar potential in India (from Ansari et al., 2013)

Land area of India	3,287,590 sq. km
Approximate number of sunny days	200
Unit potential of solar power from 1 sq. m	4 kWh/day
If conversion efficiency of solar PV cells	15%
Potential units of solar power from 1 sq. km	120 million units per year
If 0.5% of land is used for solar power installations	16,438 sq. km
Potential units of solar power from 0.5% of land	1972 billion units per year

Taking stock of the JNNSM's first phase (2010-13), one recent report was positive in its summation:

"The JNNSM Phase I (2010-13) implementation has witnessed appreciable scaling up of solar capacities in India in a short time span of three years. In addition, several state governments have declared their own state level solar policies to promote solar generation. As a result, the installed capacity of solar energy has increased from a mere 2 MW in 2008-09 to more than 1,686 MW by March 2013. The aggressive participation from the private sector in the grid-connected segment under Phase I of JNNSM has already resulted in lowering of solar tariffs for both the solar thermal and solar PV projects"[6].

The JNNSM's mandate was to promote ground-mounted as well as rooftop SPV projects. Phase 1 of the Mission saw launch of Rooftop PV and Small Scale Solar Generation Program with a view to develop both rooftop and ground-mounted PV projects with a maximum capacity of 2 MW, incentivized through GBI (generation-based incentives) linked tariffs. A report prepared by the Forum of Regulators in 2013 determined that capacity exceeding 90 MW was installed under this scheme during Phase1. However, the bulk of this capacity derived from the ground-mounted segment, with the rooftop segment receiving a "negligible" response. In general, this bias has been echoed by state solar policies and programs, although there are certainly exceptions to the rule. Overall, it can be concluded that although high-visibility national initiatives like JNNSM and state solar policies and incentive schemes have been successful in jump-starting solar-based distributed generation projects, the rooftop segment (in particular small-scale projects on the scale of the typical independent urban household) has lagged and is yet to take off. The positive side of this is lag is that there is immense room for growth in rooftop solar PV market segment [7].

In India as elsewhere in the world, the push to modernize the electricity infrastructure (the "grid") to make it more resilient and environmentally friendly has been a major source of inspiration for green thinking, and solar power has figured prominently in plans for diversifying distributed generation. India's ambitious Smart Grid Vision and Road-map, announced by the Ministry of Power in 2013, envisaged policies mandating roof top solar PV installations for large foot print establishments, i.e., those with a connected load of more than 20kW or "otherwise defined threshold" [8]. Reinforcing while also simultaneously amplifying the spatial scope of this "green" thinking, India's Ministry of New & Renewable Energy "Grid Connected SPV Rooftop, Solar Cities and Green Buildings Division" program represents an attempt to fast-track development of sixty towns/cities (designated Solar Cities), with at least one in every state, with financial support from the MNRE [9].

Against this backdrop, we report results from a demand-side study of home owners' attitudes toward, interest in and knowledge of rooftop solar in the city of Puducherry in the Union Territory of Pondicherry. Puducherry has been the site of a successful MoP Smart Grid pilot designed and implemented by Power Grid Corporation of India Limited (PGCIL) in collaboration with the Puducherry Electricity Department (PED) [10]. Puducherry falls under the purview of the MNRE's ambitious Solar City program. For our preliminary study, we surveyed 50 predominantly private, independent home-owners residing in areas covered by the PGCIL-PED Smart Grid pilot. Our focus here is roof-top solar PV system. Where applicable, we contextualize the survey responses with data from oral, side-bar interactions with respondents as they completed the survey in our (physical) presence. These interactions provided a complementary source of rich data. Respondents' questions on and concerns over financial details on the technology provided valuable substantive additional information, for example, which we found valuable in interpreting the survey data.

This chapter is organized as follows. As background, in the following section we outline two solar PV installation and metering configurations. We then contextualize our survey study with a look at distributed generation in Puducherry and solar-relevant provisions in the recently-concluded Phase-1 of the PGCIL Smart Grid pilot before presenting the survey results. We conclude with a discussion of the findings.

Solar PV System, Metering and Ownership Configurations

Two commonly available configurations are rooftop solar PV system with and without battery back-up as shown in Figures 1 and 2.

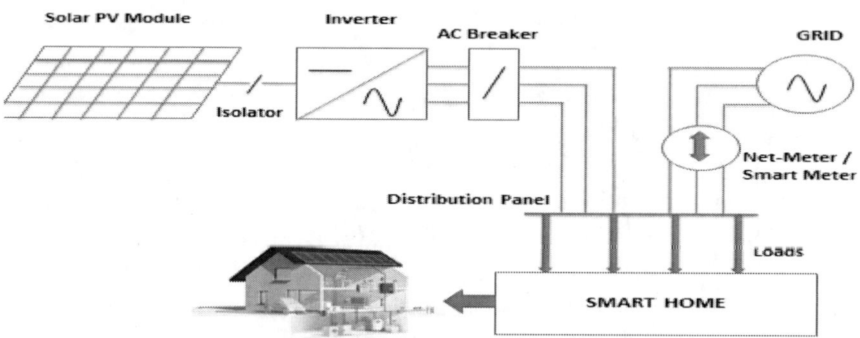

Figure 1. SPV System without Battery Back-up.

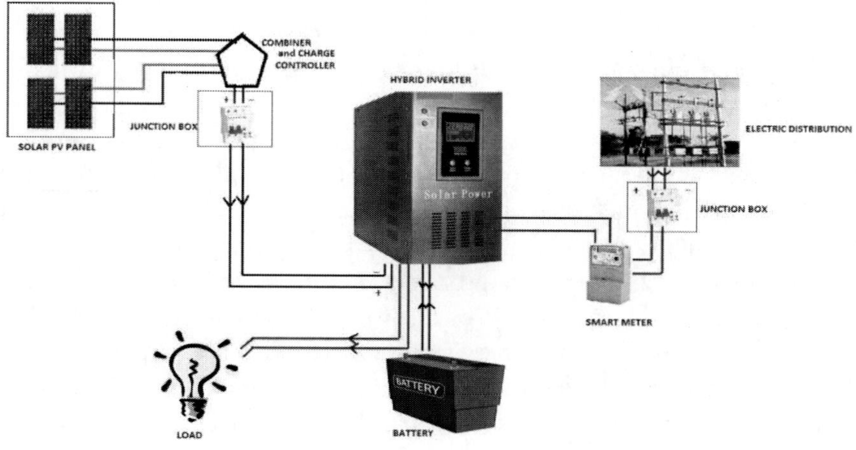

Figure 2. SPV System with Battery Back-up.

Customer cost profiles are different for these two options. The estimated cost of a 1 kW rooftop solar installation is Rs. 1.95 lakh with battery back-up and Rs. 1.67 lakh without [11]. For Puducherry residents, subsidies up to 50% of the cost may be available from MNRE and the Puducherry Government. The customer stands to benefit from lower power cost as well: "Cost of electricity generation would be Rs. 3.85/Unit and Rs. 2.75/Unit for the Rooftop Solar Power System with battery & without battery respectively. This cost shall be for 25 years life span and would not increase, which provide better option against conventional power priced around Rs. 5 per unit and increasing every year" [11].

Two metering configurations are possible: gross and net metering. With gross-metering, the entire energy output from the rooftop solar system is fed directly into the electrical grid. The system owner benefits from selling solar power to the utility at the feed-in-tariff (FiT) rate. With the net metering configuration, the system owner installs the system primarily to self-consume the power that is generated. The power that is not self-consumed can be banked or sold to the utility. The net-meter is a bi-directional meter which can register both import of power from the grid and export of power to the grid. Capital subsidies, generation-based incentives and tax credits are used to incentivize the net metering arrangement, while FiT is used with gross metering [6]. Both options have been implemented in India and elsewhere.

Two ownership options are available for rooftop solar PV installations: self-owned and third-party owned (leased). In the self-owned, net-metered arrangement, the net power consumed – that is, the difference between the units of power imported from the grid and units exported to the grid from the solar PV system – is credited to the owner's account and may be used to offset (future) power imports from the grid. In a third-party owned, net-metered set up, the owner of the rooftop leases the solar PV system from a third-party provider for a monthly rental fee. The rooftop owner uses power generated from the system while exporting any excess to the grid. Both arrangements have their advantages. For power consumers, rooftop systems serve as a source of power for captive consumption even when the grid is unavailable. They have the opportunity to be prosumers – producers as well as consumers of power, with attendant financial benefits. The third party-owned option protects the rooftop owner from high upfront installation costs and technology risk.

Power System Infrastructure for Electric Vehicle in India an Overview

Introduction

The electrical power distribution system of India is bulky, complex and growing rapidly. The 8% growth rate of GDP would lead to the rise in demand by 3 times in next decade and 66% of which would be on-grid only [1]. By 2032, the expected demand of the Power would be 900GW [2-3], and this demand would be met significantly by renewable sources besides the conventional resources. The expected potential of renewable energy would be 183GW by the same time [4-5]. Along with the availability of power, the reliability and efficiency also become prime concerns for the Indian power systems. The increase in demand would not only consist of conventional domestic and industrial type loads, but also few new unique loads like electric vehicle and power electronics loads which would contribute to a major part in demand growth.

As per the national mission for electric mobility, India would be having six million electric vehicles on road by 2020 in [6-7]. Further, in current scenario, distribution system in India is facing high Aggregate Technical and Commercial (AT&C) losses [8-9]. According to report from the International Energy Agency (IEA) on Transmission and Distribution (T&D) losses in different countries during year 2010-11, the T&D losses in India are 23.65% against 9.8% average throughout the world [10]. The solution for all the above problems is to have a smart electric grid at distribution level. With respect to a conventional grid, smart grid is more robust, transparent, reliable and efficient. Besides, the smart grid is also beneficial in terms of cost reduction, improvement in power quality, increase in national productivity, safety and also the direct participation of consumers. Smart grid has become the next level of power system worldwide [11].

To evaluate the real benefits and to identify suitable technologies/models of the smart grid, Ministry of Power, Govt. of India proposed 14 pilot projects across the country with different functionalities of smart grid. Currently, all these pilot projects are in the initial stage of implementation. The main objectives of these pilots are indigenization of technology, development of scalable and replicable models, bringing up suitable standards and regulations based on experience of these pilot projects. Puducherry smart grid project [12] is one of the proposed pilots which is being developed jointly by Power Grid

Corporation of India Limited (POWERGRID) along with open collaborators and Puducherry Electricity Department (PED).

In smart grid projects, we have conventional sources, solar panels, wind generators, storage devices, electric vehicles, domestic loads, industrial loads and various other interfacing devices. All these components have their own characteristics of operations which may lead to deformation in power quality if their operation is not monitored and controlled properly. Further, the varieties of loads as well as interfacing devices inject harmonics into the system which deteriorates the power quality. The introduction of more power electronic devices gives an increase in harmonic distortion [13]. Further the PV integration into the network influences both the voltage and network losses positively [14, 15]. Power Quality (PQ) disturbances include those from short to long duration variations, harmonics, flickers, increased downtime, etc. [12, 16] and poor power quality would result in incurring high operation expenditure (Opex) cost. By monitoring and analyzing power quality, the cause of power system disturbance can be identified and improved before they cause interruptions [13, 17]. Various standards like IEEE-1547, IEEE-519, IEEE-1159 etc. are laid down to monitor and control the quality of power supply. The indexes to monitor power quality are frequency variations, voltage variations, harmonics, flicker, power factor, etc. [18, 19].

Puducherry smart grid pilot project covers Division-1 of ten divisions in PED. With 87035 numbers of consumers, Division-1 has 100% electrification. About 79% of these consumers are domestic and the rest includes commercial, HT, agriculture, streetlight, etc. The entire division is supplied by one 110/22/11 kV sub-station, which feeds seven 22kV overhead feeders, five 11 kV underground cable feeder and 325 distribution transformers with total load of 127.8 MVA. The completed interim pilot project covers 1400 consumers in 22 kV town feeder with nine Distribution Transformers (DTs). Pilot project showcases all features of smart grid like Advance Metering Infrastructure (AMI), Power Quality Management (PQM), Peak Load Management (PLM), Outage Management System (OMS), Renewable Energy Integration and Energy Storage [12].

In this Chapter, the power quality of the Pondicherry smart grid pilot project is investigated. The major contributions of the paper are as follows:

- Extensive field data are collected and analyzed to see the power quality issues.
- The existing PQ problems are compared with the standards.

- Total harmonic distortion, flicker, harmonic subgroups, individual harmonics components are obtained.

Due to various power electronics load and interfaces, it is observed that there are some concerns in few areas. It was found that there is already an active power filter in the system for improving the power quality of the supply. The results are very encouraging and can be used for many smart grid projects.

Power Quality Components

Power quality is affected by several factors [18, 20] as discussed below.

Frequency Variation

The frequency of power supply in India is 50Hz. To maintain the power quality, the Central Electricity Regulatory Commission (CERC) India allows the variation from 49.7 to 50.2 Hz. Table 1 also presents the various standards for the same.

Voltage Variation

The voltage variation refers to the variation in voltage from its mean value. As per standards, only 10% variation is acceptable to maintain the quality of supply. Variation beyond 10% would be referred to as poor quality.

Table 2. Power Quality Index

PQ Indicators	Voltage levels	Standards		
		IEC61000	EN50160	IEEE
Frequency Variation	-	±1%	±1%	-
Voltage Variations	LV & MV	±10%	±10%	-
	HV & EHV	±10%	-	-
Interruption	all	±10%	±1%	±10%
Imbalance	LV & MV	±2%	±2%	-
	HV & EHV	±1%	±1%	-
Harmonics	LV	(8,6,2)	(8,6,2)	(5,3,3)
	MV	(6.5,5,1.6)	(8,6,2)	(5,3,3)
	HV	(3,2,1.5)	-	(2.5,1.5,1.5)
	EHV	(3,2,1.5)	-	(1.5,1,1)
Flicker	LV	(1,0.8)	(-, 1)	-
	MV	(0.9,0.7)	(-, 1)	-
	HV & EHV	(0.8,0.6)	-	-

Table 3. Voltage harmonics limit as per standard EN50160

Odd Harmonics		Event Harmonics	
Harmonic Rank	Harmonic Voltage [%]	Harmonic Rank	Harmonic Voltage [%]
3	5	2	2
5	6	4	1
7	5	6	0,5
9	1,5	8	0,5
11	3,5	10	0,5
13	3	12	0,2

Voltage Distortion

Voltage distortion can be understood as loss of sinusoidal nature of the waveform and is calculated by Total Harmonic Distortion (THD) and individual harmonics up to 50th order. The standards for harmonics can be seen in tables above.

Harmonics and Inter-Harmonics Subgroup

Apart from individual harmonic, the harmonic and inter-harmonic subgroups also contribute to power quality analysis. The harmonic subgroup includes $\pm 5Hz$ as shown in Figure 3 and inter-harmonic subgroup includes the frequencies between two harmonic subgroups [21].

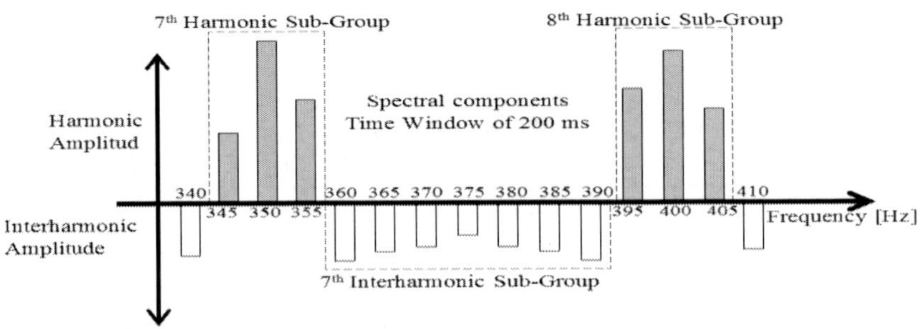

Figure 3. Harmonics and Inter-harmonics subgroup.

Flicker

RE generators are mostly interfaced through Convertor-Invertor. Power Electronics inject Harmonics into Grid causing flicker pollution etc., Voltage Flicker or Current Flicker. Use of Harmonic Filters reduces Flicker and through selection of Passive, Active, Hybrid which are to be done properly. Soft starting mechanism for Induction WTG can reduce Voltage Harmonics as seen from Figures 4 & 5.

The flicker calculation includes both short-term (PST) and long-term (PLT) flickers. The limit of flicker can be obtained from the Table 1.

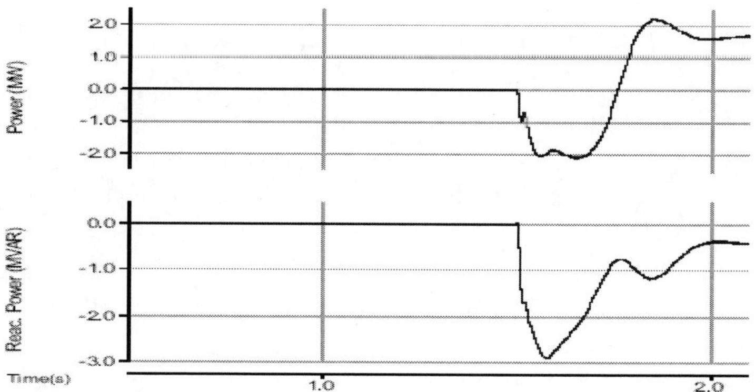

Figure 4. Voltage Flicker during Startup.

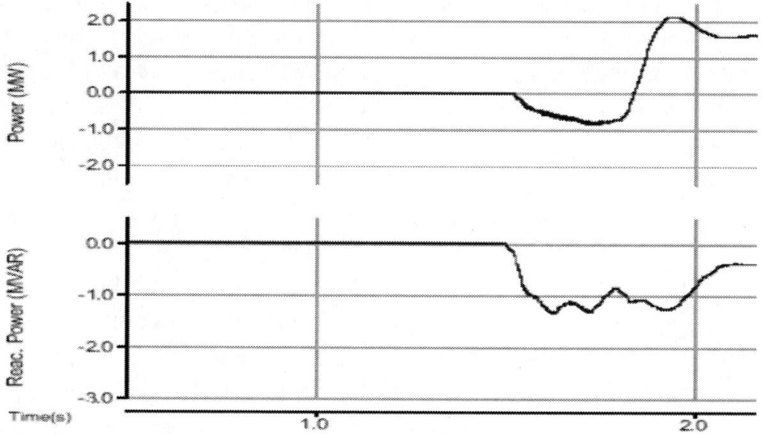

Figure 5. Voltage Flicker with Soft start.

Hybrid Electric Vehicles

When considering lifetime CO2 emissions, the convergence of renewable energy will make EVs cleaner than conventional fuel vehicles. Distributed solar energy will help reduce transmission and distribution losses, lowering lifetime CO2 emissions as well as the operating costs of EVs and accelerating their commercial viability. When lifetime CO2 emissions considered for the convergence of green sources will make EVs smoother than internal combustion vehicles, distributed solar energy will aid in the reduction of losses associated with distribution and transmission, reducing lifetime Dioxide (CO2) emissions as well as EV operating costs and accelerating commercial viability.

The automobile industry has a high level of power or fuel consumption. Thanks to whoever came up with the concept of utilising solar energy in hybrid vehicles, photovoltaic electric cars are a cost-effective alternative. A photovoltaic solar vehicle is a type of road transport. They have been successfully carried by renewable energy, such as solar power used to recharge batteries. Solar panels are integrated into the body of the rechargeable battery, making it self-sufficient. Electric vehicles use electricity for charging their battery cells rather than fossil fuels such as gasoline or diesel. With more EV battery charging points springing up, to boost their battery cells at the nearest station rather than endure lengthy queues at CNG stations or petrol pumps. Vehicle owners can also start charging their batteries from the home environment in which they live using charging equipment.

There will be an initial charging period, after which the storage power is ready to supply enough power for the vehicle to travel. They effectively decrease smog and provide encouraging growth in solar-powered fuel savings. Solar panels are integrated into the body of the rechargeable battery, making it self-sufficient. There will be an initial charging period, after which the storage power is ready to supply enough power for the vehicle to travel. They effectively decrease smog and provide encouraging growth in solar-powered fuel savings.

Solar energy is a low-cost solution with a 25-30-year lifespan. Because electric vehicles have fewer moving parts than internal combustion vehicles, they require less maintenance. Electric vehicles require less maintenance than conventional gasoline or diesel vehicles. It helps to make their energy bills more effective while lowering your costs. Solar system maintenance and monitoring are inexpensive. As a result, the annual cost of operating an electric vehicle is dramatically lower. Because there is no engine under the undercarriage, electric vehicles can operate in absolute silence. There will be no noise due to the absence of the propulsion system. The electric motor runs so quietly that it's impossible to peek into your instrument cluster to confirm that it's converted forward.

The power-train (PT) controller has played a significant role in the photovoltaic battery powered hybrid electric vehicle (HEV). The primary goal of the PT controller is to ensure a battery management system with good load regulation and maximum power extraction from photovoltaic panels whenever possible. The power-train controller is divided into two levels: lower-level controllers and a high-level control algorithm. Lower-level controllers are designed to handle specific tasks like maximum power point tracking, battery charging, and load regulation. The maximum power point tracking algorithm based on perturb and observe is used to extract maximum power from solar photovoltaic panels, while the battery charging controller is designed using a PI controller. The overall efficiency of the Electric vehicle can be enhanced by boosting the effectiveness of the power-train's individual elements. Battery health is among the most significant variables determining the overall efficiency of the EV power-train system. Battery management systems (BMS) can extend the battery's life and health. Even though Vehicles have more electrically powered charging and discharging cycles, especially plug - in electric vehicle-to-grid (V2G) and grid-to-vehicle (G2V), and since renewable resources are intermittent, batteries will experience several inadequate and temporary charge/discharge cycles, limiting their lifetime.

Solar powered hybrid electric vehicles are not affordable to everyone due to their high cost. It is not inaccurate to say that electric vehicles are only obtainable to the wealthy. One of the primary reasons for its high cost is that the number of electric cars availability is restricted due to product accessibility. Low resource costs and rising prices are expensive.

To conclude, electric vehicles have both advantages and disadvantages. They are an excellent way to reduce environmental pollution but have some drawbacks. Electric vehicles are not affordable to everyone due to their high cost. It is not inaccurate to say that electric vehicles are only available to the wealthy. One of the primary reasons for its high price is that the range of electric cars available is also restricted due to product usage. Low financial asset expenses and rising prices are expensive.

References

[1] M. Wang, The Role of Charging Infrastructure in the Adoption of Electric Vehicles, *Journal Transportation Research Part D: Transport and Environment*, 2020.

[2] Dale Hall, Nic Lutsey, *Emerging Best Practices for Electric Vehicle Charging Infrastructure, ICCT*. October (2017).

[3] Philipp Borgstedt, Bastian Neyer, Gerhard Schewe, Paving the road to electric vehicles – A patent analysis of the automotive supply industry, *Journal of Cleaner Production*, 167, 2017, 75-87.

[4] Dale Hall, Marissa Moultak, Nic Lutsey, *Electric Vehicle Capitals of the World - Demonstrating the Path to Electric Drive, ICCT*. March (2017).

[5] C. Angelo Guevara, Esteban Figueroa, Marcela A. Munizaga, Paving the road for electric vehicles: Lessons from a randomized experiment in an introduction stage market, *Transportation Research Part A: Policy and Practice*, 153, 2021, 326-340.

[6] Haonan Xie, Renhao Huang, Hui Sun, Zepeng Han, Meihui Jiang, Dongdong Zhang, Hui Hwang Goh, Tonni Agustiono Kurniawan, Fei Han, Hui Liu, Thomas Wu, Wireless energy: Paving the way for smart cities and a greener future, *Energy and Buildings*, 297, 2023, 113469.

[7] Sulabh Sachan, Sanchari Deb, Sri Niwas Singh, Different charging infrastructures along with smart charging strategies for electric vehicles, *Sustainable Cities and Society*, 60, 2020, 102238.

[8] López, G., J. I. Moreno, H. Amarís, F. Salazar, Paving the road toward Smart Grids through large-scale advanced metering infrastructures, *Electric Power Systems Research*, 120, 2015, 194-205.

[9] Kenneth Holmberg, Ali Erdemir, The impact of tribology on energy use and CO2 emission globally and in combustion engine and electric cars, *Tribology International*, 135, 2019, 389-396.

[10] Paaso, D. Kushner, S. Bahramirad and A. Khodaei, Grid Modernization Is Paving the Way for Building Smarter Cities [Technology Leaders], *in IEEE Electrification Magazine*, 6, 2018, 6-108.

[11] Ghazale Haddadian, Mohammad Khodayar, Mohammad Shahidehpour, Accelerating the Global Adoption of Electric Vehicles: Barriers and Drivers, *The Electricity Journal,* 28, 2015, 53-68.

[12] Mei, Shengwei and Li, Rui and Xue, Xiaodai and Chen, Ying and Lu, Qiang and Chen, Xiaotao and Ahrens, Carsten D. and Li, Ruomei and Chen, Laijun, Paving the way to smart micro energy grid: concepts, design principles, and engineering practices, *CSEE Journal of Power and Energy Systems,* 3. 2017, 440-449.

[13] Ran Tao, Chi-Wei Su, Bushra Naqvi, Syed Kumail Abbas Rizvi, Can Fintech development pave the way for a transition towards low-carbon economy: A global perspective, *Technological Forecasting and Social Change*, 174, 2022, 121278.

[14] Stefan Metz, Linde, Pioneers hydrogen compression techniques for fuel cell electric vehicles, *Fuel Cells Bulletin,* 2014, 2014, 12-15.

[15] Eftekhari, Ali, Lithium Batteries for Electric Vehicles: From Economy to Research Strategy, *ACS Sustainable Chemistry & Engineering, American Chemical Society,* 7(6), 2019, 5602 – 5613.

[16] Georgina Santos, Road transport and CO_2 emissions: What are the challenges? *Transport Policy,* 59, 2017, 71-74.

[17] Danese, A., Torsæter, B. N., Sumper, A., Garau, *M. Planning of High-Power Charging Stations for Electric Vehicles: A Review. Appl. Sci.* (2022) *12*, 3214.

[18] Zhou Bochao, Pei Jianzhong, Calautit John Kaiser, Zhang Jiupeng and Guo Fucheng, *Solar self-powered wireless charging pavement—a review on photovoltaic pavement and wireless charging for electric vehicles, Sustainable Energy Fuels*, VL - 5, IS - 20, The Royal Society of Chemistry (2021).

[19] Faran Razi, Ibrahim Dincer, *A review of the current state, challenges, opportunities and future directions for implementation of sustainable electric vehicle infrastructure in Canada, Journal of Energy Storage*, Volume 56, Part B (2022),106048.

[20] Kore, Hemant Harishchandra;Koul, Saroj, *Electric vehicle charging infrastructure: positioning in India, Management of environmental quality: An international journal*, 10 Mar (2022) pages 776 – 799.

[21] Sina Ibne Ahmed, Hossein Salehfar and Daisy Flora Selveraj, *Grid Integration of PV Based Electric Vehicle Charging Stations: A Brief Review,* North American Power Symposium (NAPS), 9-11 (Oct. 2022).

Chapter 2

Hybrid Electric Vehicles (HEVs): Transforming Modern Transportation

D. Menaga*

Department of Electrical and Electronics Engineering,
Saveetha School of Engineering,
Saveetha Institute of Medical and Technical Sciences, Saveetha University,
Chennai, Tamil Nadu, India

Abstract

Hybrid Electric Vehicles (HEVs) are a significant development in the automotive industry, offering numerous advantages over conventional internal combustion engine vehicles. HEVs combine two or more power sources to enhance efficiency, reduce emissions, and optimize energy consumption. They can operate in various modes, seamlessly shifting between the internal combustion engine, electric motor, or a combination of both based on driving conditions and energy requirements. HEVs come in various configurations, each with its own advantages and use cases. Parallel hybrids feature both an internal combustion engine and an electric motor that are mechanically connected to the transmission, allowing both to simultaneously power the wheels. Series hybrids use the internal combustion engine exclusively as a generator to power an electric motor that drives the wheels. Plug-In Hybrid Electric Vehicles (PHEVs) incorporate a larger battery that can be charged via an electrical outlet, enabling extended electric-only driving. Mild hybrids feature a smaller electric motor and battery system that predominantly serves to assist the internal combustion engine, especially during acceleration and stop-start driving situations. Hydrogen Fuel Cell Hybrid Vehicles deploy

---

* Corresponding Author's Email: menaganit@gmail.com.

In: Electric Vehicle Technology Structure, Instrumentation and Challenges
Editors: S. N. Sundaram, P. N. Sundaram, M. H. Kumar et al.
ISBN: 979-8-89113-695-3

a hydrogen fuel cell stack to generate electricity that powers an electric motor.

Keyword: Hybrid Electric Vehicles, fuel efficiency, emissions reduction, sustainable transportation, electric-only driving

Introduction

In recent years, the automotive industry has undergone a remarkable transformation, primarily driven by the urgent need to reduce greenhouse gas emissions, combat air pollution, and transition toward more sustainable transportation methods. Central to this shift is the emergence of Hybrid Electric Vehicles (HEVs), which blend conventional internal combustion engines with electric power to create more efficient and environmentally friendly vehicles. In this comprehensive exploration, we will delve into the world of HEVs, beginning with a detailed definition and an analysis of the various types of HEVs in [1]. We will then assess the significance of HEVs in modern transportation, covering their influence on fuel efficiency, emissions reduction, their role in the transition to sustainable transportation, and their adaptation to evolving regulatory frameworks. Hybrid Electric Vehicles (HEVs) are a significant development in the automotive industry, offering numerous advantages over conventional internal combustion engine vehicles. HEVs combine two or more power sources to enhance efficiency, reduce emissions, and optimize energy consumption in [2]. They can operate in various modes, seamlessly shifting between the internal combustion engine, electric motor, or a combination of both based on driving conditions and energy requirements.

Definition and Types of HEVs

Definition of HEVs

Hybrid Electric Vehicles, or HEVs, represent a paradigm shift in the automotive industry. These vehicles leverage two or more power sources to enhance efficiency, curtail emissions, and optimize energy consumption. Typically, HEVs combine an internal combustion engine (typically gasoline or diesel) with an electric motor and a rechargeable battery. The defining feature of HEVs is their ability to operate in various modes, seamlessly shifting between the internal combustion engine, electric motor, or a combination of both based on driving conditions and energy requirements. This intricate coordination of power sources is orchestrated by advanced control systems, rendering HEVs a pinnacle of automotive engineering.

Types of HEVs

HEVs manifest in diverse configurations, each offering unique advantages and use cases. Here are some common types of HEVs:

Parallel Hybrid

Parallel hybrid vehicles feature both an internal combustion engine (ICE) and an electric motor, mechanically connected to the vehicle's transmission. This mechanical linkage allows both the ICE and the electric motor to simultaneously provide power to the wheels. Parallel hybrids are capable of running solely on the electric motor, the gasoline engine, or a hybrid mode that combines both power sources. This versatility empowers parallel hybrids to adapt to a variety of driving scenarios.

Series Hybrid

Series hybrid vehicles, in contrast, use the internal combustion engine exclusively as a generator. In this setup, the gasoline engine's primary role is to generate electricity, which is then utilized to power an electric motor responsible for driving the wheels. Importantly, the internal combustion engine does not directly transmit mechanical power to the wheels. Series hybrids are known for their proficiency in converting gasoline into electrical energy, which can be particularly advantageous in situations requiring steady-state power generation.

Plug-In Hybrid (PHEV)

Plug-In Hybrid Electric Vehicles, or PHEVs, represent an evolution in hybrid technology. They incorporate a larger battery compared to non-plug-in hybrids, and this battery can be charged via an electrical outlet [3]. The expanded battery capacity enables PHEVs to offer an extended electric-only driving range. While traditional hybrids rely on regenerative braking and the internal combustion engine for battery replenishment, PHEVs can be charged externally, such as through a home charging station or a public charging point. This feature is especially beneficial for urban commutes, enabling prolonged electric-only operation and consequently reducing gasoline consumption and emissions.

Mild Hybrid

Mild hybrid vehicles feature a smaller electric motor and battery system compared to full hybrids. In a mild hybrid configuration, the electric motor predominantly serves to assist the internal combustion engine, especially during acceleration and stop-start driving situations. While these hybrids may not deliver extended electric-only driving capabilities, their mild hybrid systems enhance overall fuel efficiency and diminish emissions, rendering them a practical choice for city driving conditions.

Hydrogen Fuel Cell Hybrid

Hydrogen Fuel Cell Hybrid Vehicles constitute a niche category of HEVs that harness hydrogen fuel cell technology. These vehicles deploy a hydrogen fuel cell stack to generate electricity, which powers an electric motor responsible for propelling the vehicle. One notable characteristic of hydrogen fuel cell hybrids is their near-zero emissions profile, as they generate only water vapor and heat as byproducts. This makes them an ecologically friendly option, provided that the hydrogen used is produced through clean and sustainable methods. Although hydrogen fuel cell technology is still in the developmental stages, it holds promise as a zero-emission alternative in the automotive sector.

Each type of HEV is engineered to meet specific driving needs and operational conditions. Their distinctive powertrain configurations enable automakers to cater to a wide range of consumers with diverse preferences and requirements.

Significance of HEVs in Modern Transportation

The significance of Hybrid Electric Vehicles in the context of modern transportation is pivotal and multifaceted. These innovative vehicles encompass various facets of sustainability, efficiency, and adaptability. The following are key factors that underscore the importance of HEVs in today's automotive landscape:

Fuel Efficiency

One of the foremost rationales behind the development of HEVs is the imperative to enhance fuel efficiency. Traditional internal combustion engine vehicles inherently suffer from inefficiencies, with substantial energy losses occurring during processes like idling, deceleration, and low-speed driving. HEVs, conversely, incorporate advanced technologies that optimize energy usage. Their ability to function in electric-only mode, particularly during low-speed urban driving, substantially curbs fuel consumption [4]. Furthermore, they integrate regenerative braking systems that capture and store energy typically dissipated as heat in conventional vehicles. This stored energy can be redeployed to aid in propulsion, further augmenting fuel efficiency. Consequently, HEVs are acclaimed for their capacity to attain higher miles per gallon (MPG) and markedly reduce the overall cost of vehicle ownership through diminished fuel consumption.

Emissions Reduction

HEVs play a pivotal role in abating environmental issues linked to air pollution and greenhouse gas emissions. Internal combustion engines, especially those powered by gasoline and diesel, release pollutants such as carbon dioxide (CO2), nitrogen oxides (NOx), and particulate matter [5]. These emissions contribute to smog formation, climate change, and adverse health effects. In contrast, HEVs, when operating in electric mode, yield zero tailpipe emissions. Even when the internal combustion engine is engaged, it operates more efficiently, resulting in lower emissions per mile driven [6]. HEVs, thus, constitute a significant stride toward mitigating the ecological footprint of the transportation sector.

Transition to Sustainable Transportation

HEVs serve as a pivotal element in the ongoing shift toward sustainable and environmentally responsible modes of transportation. They serve as a bridge between traditional gasoline and diesel vehicles and fully electric vehicles. While pure electric vehicles (EVs) are increasingly gaining traction, they may not be feasible or accessible for all consumers due to concerns related to driving range, charging infrastructure, and initial acquisition costs **[7].** HEVs provide a pragmatic solution by delivering the advantages of electrification without the constraints of range anxiety. They can function in hybrid mode, employing a blend of internal combustion and electric power, making them versatile and capable of long-distance travel. This transitional role permits consumers to adapt to electric vehicle technology and benefit from improved efficiency and reduced emissions without the full commitment to EV infrastructure.

Enhanced City Driving

Urban areas are frequently plagued by traffic congestion, noise pollution, and air quality issues. HEVs are ideally suited for city driving, where frequent stops and slow-moving traffic are routine. When operating in electric mode, HEVs generate minimal noise and zero tailpipe emissions, contributing to a quieter and cleaner urban environment. In densely populated cities, where mitigating pollution and noise is imperative, HEVs offer an elegant solution by reducing environmental impact and enhancing the quality of life for city residents.

In conclusion, Hybrid Electric Vehicles have become a linchpin in the ongoing transformation of the automotive industry. They provide a practical and eco-friendly solution to the challenges of fuel efficiency and emissions reduction while paving the way for more sustainable modes of transportation in the future. As technology continues to advance, HEVs are likely to play an increasingly significant role in modern transportation systems.

Battery Technology Advancements in Hybrid Electric Vehicles (HEVs)

Evolution of HEV Battery Technology

The evolution of battery technology is central to the advancement and widespread adoption of Hybrid Electric Vehicles (HEVs). HEVs rely on high-performing batteries to store and deliver electrical energy efficiently. This section explores the historical development and technological progression of HEV battery technology.

Early Battery Technologies

The early years of HEVs saw the utilization of nickel-metal hydride (NiMH) batteries as the primary energy storage solution. These NiMH batteries, while more energy-dense than traditional lead-acid batteries, had limitations in terms of energy capacity and power density. They were a significant improvement over lead-acid batteries, but they were still relatively heavy and occupied a substantial amount of space within the vehicle.

Advancements in NiMH Batteries

As HEV technology matured, advancements in NiMH battery technology led to improved energy density and durability. These batteries became more compact and offered higher energy-to-weight ratios, which allowed for greater storage capacity and reduced overall weight. These improvements enhanced the efficiency and performance of HEVs, enabling longer electric-only driving ranges and better fuel economy.

Transition to Lithium-Ion Batteries

The transition from NiMH to lithium-ion (Li-ion) batteries marked a significant turning point in HEV battery technology. Li-ion batteries are known for their higher energy density, power density, and overall performance. They are lighter and more compact than NiMH batteries, offering automakers greater flexibility in designing HEV powertrains.

Advancements in Li-Ion Batteries

Advancements in Li-ion battery technology have continued to improve the efficiency and practicality of HEVs. These advancements include:

- Energy Density: Li-ion batteries have seen a steady increase in energy density, allowing them to store more energy in the same physical space. This directly translates into extended electric-only driving ranges for PHEVs and improved energy utilization for conventional HEVs.
- Power Density: Higher power density means that Li-ion batteries can deliver more power in a shorter amount of time. This is crucial for providing quick bursts of power when acceleration is needed, contributing to the overall performance of the vehicle.

- Cycle Life: Modern Li-ion batteries are designed to have a longer cycle life, which means they can undergo more charge and discharge cycles before their capacity significantly degrades. This longevity is essential for the durability and reliability of HEVs.
- Fast Charging: Li-ion batteries are compatible with fast-charging technology, reducing charging times significantly. This feature is particularly valuable for PHEVs, as it allows drivers to recharge their batteries quickly and conveniently.
- Thermal Management: Advanced thermal management systems in Li-ion batteries help regulate temperature, ensuring safe and efficient operation. Effective thermal management is critical for prolonging battery life and preventing overheating.
- Cost Reduction: Over the years, improvements in battery manufacturing processes and economies of scale have led to cost reductions, making HEVs more affordable for consumers.

Solid-State Batteries

The future of HEV battery technology holds the promise of solid-state batteries. Solid-state batteries are seen as the next frontier in energy storage. They offer numerous advantages, including even higher energy density, enhanced safety, and the potential for longer life cycles. Furthermore, solid-state batteries have the potential to overcome some of the limitations associated with liquid electrolytes, which can be prone to overheating and fire risk.

As these advanced batteries continue to develop, they are expected to have a substantial impact on the performance, efficiency, and affordability of HEVs. Solid-state batteries could enable even longer electric-only driving ranges and faster charging times, further enhancing the appeal of HEVs.

Cost Reduction and Widespread Adoption

The cost of batteries has long been a significant factor influencing the affordability and widespread adoption of HEVs. Battery packs can account for a substantial portion of the vehicle's overall cost. As battery technology has advanced, several factors have contributed to cost reduction and the broader acceptance of HEVs.

Economies of Scale

One of the primary drivers of cost reduction has been the increasing production volumes of HEVs. As more HEVs are manufactured, the cost of producing battery packs decreases due to economies of scale. Larger production runs lead to cost efficiencies in manufacturing processes, material sourcing, and assembly.

Improved Manufacturing Techniques

Advancements in battery manufacturing techniques have streamlined the production process, making it more efficient and cost-effective. Techniques like automated assembly and precise quality control have contributed to reducing production costs.

Battery Recycling

Another element in cost reduction is the development of battery recycling programs. As HEV batteries reach the end of their useful life in vehicles, they can still hold a significant amount of energy storage capacity. By recycling and repurposing these batteries for stationary energy storage applications, the overall cost of battery production is reduced, and the environmental impact is minimized.

Research and Development

Investments in research and development have led to innovative solutions for enhancing battery performance while reducing costs. This includes the exploration of alternative materials, improved manufacturing processes, and new battery chemistries.

Incentives and Subsidies

Many governments and regions have implemented incentives, tax credits, and subsidies to promote the adoption of HEVs. These financial incentives can significantly reduce the upfront cost of purchasing an HEV, making them more attractive to consumers.

Competitive Market

The competitive market for HEVs has pushed manufacturers to innovate and reduce costs. This competition benefits consumers by offering a wider range of affordable HEV options.

As battery technology continues to advance and manufacturing costs decrease, HEVs are becoming more accessible to a broader range of consumers. With these cost reductions, HEVs are positioned to play an even more significant role in the transition to cleaner and more sustainable transportation.

In summary, the evolution of battery technology within HEVs has been a critical factor in their growth and acceptance in the automotive industry. Advancements in energy density, power density, cycle life, fast charging, and thermal management have improved the performance and appeal of HEVs. Moreover, cost reductions driven by economies of scale, improved manufacturing techniques, recycling, research and development, and government incentives have made HEVs more affordable and accessible to consumers. With the promise of solid-state batteries on the horizon, the future of HEVs appears brighter than ever, with the potential for even greater efficiency and performance.

Plug-In Infrastructure Development: Charging Infrastructure for PHEVs

Growth of Charging Infrastructure for PHEVs

The surge in Plug-In Hybrid Electric Vehicles (PHEVs) is a significant facet of the global push toward cleaner and more sustainable transportation [3]. As PHEVs become increasingly popular among consumers, the expansion of charging infrastructure has become essential in facilitating their adoption. This comprehensive exploration delves into the critical components of the growth of charging infrastructure for PHEVs, discussing its implications, government and industry initiatives, and its profound impact on PHEV adoption.

Charging Station Networks

One of the most conspicuous manifestations of the growth of charging infrastructure for PHEVs is the emergence of extensive charging station networks. These networks encompass a variety of charging stations, including Level 1, Level 2, and DC fast chargers, thoughtfully positioned in urban areas, along highways, at commercial facilities, and within residential communities [4]. This diversity in charger types empowers PHEV owners to select the charging option that best aligns with their specific requirements.

- Level 1 Chargers: Level 1 chargers are fundamental chargers that rely on a standard 120-volt household electrical outlet. While their charging rate is relatively slow, they offer unparalleled convenience as they can be employed with the existing electrical infrastructure. PHEV owners can effortlessly plug in their vehicles at home, at work, or virtually anywhere with a standard outlet.
- Level 2 Chargers: In contrast, Level 2 chargers operate at 240 volts, significantly enhancing the charging rate when compared to Level 1 chargers. These chargers are prevalent in public charging stations, parking lots, and commercial facilities. Level 2 charging is notably quicker and can provide a full charge to a PHEV in a matter of hours.
- DC Fast Chargers: DC fast chargers serve as the fastest charging option for PHEVs. These chargers function at higher voltages and currents, resulting in rapid charging. They are commonly located along highways and major transportation routes, rendering long-distance travel in a PHEV more feasible.

The development of charging station networks is driven by both government and private sector investments. Government initiatives, including grants, subsidies, and regulations aimed at fostering the construction of charging infrastructure, have played a pivotal role in stimulating growth. Simultaneously, private entities such as automotive manufacturers, utilities, and charging station providers have made substantial investments in expanding the network.

Government and Industry Initiatives

Government Initiatives

Governments across the globe acknowledge the pivotal role that charging infrastructure plays in the adoption of PHEVs. To promote the expansion of charging networks, governments have enacted various initiatives:

- Financial Incentives: Governments often extend financial incentives to individuals, businesses, and charging station operators to invest in and employ charging infrastructure. These incentives can encompass subsidies, tax credits, and grants, reducing the financial burden associated with charging station deployment and utilization.
- Regulatory Requirements: Certain regions have introduced regulatory requirements necessitating the inclusion of charging infrastructure in commercial and residential buildings. These regulations encourage property developers to construct infrastructure that supports PHEVs, ensuring that charging stations are readily available.
- Research and Development Grants: Governments allocate research and development grants aimed at advancing charging technology. These grants fuel the progress of charging solutions, driving the development of cutting-edge charging technology that is faster, more efficient, and more convenient.
- Public-Private Partnerships: Collaborative efforts between governments and private entities have proven to be effective in funding, constructing, and maintaining charging networks. Public-private partnerships leverage both public resources and the expertise of the private sector, enabling the creation of comprehensive charging solutions.

Industry Initiatives

Automotive manufacturers, utility companies, and charging station providers are actively engaged in the expansion of charging infrastructure:

- Automaker Involvement: Numerous automakers recognize that the availability of charging infrastructure is critical to the success of their PHEV models. As a result, they invest in the development and deployment of charging stations and may offer charging services to their customers. This commitment not only demonstrates the automakers' dedication to sustainability but also creates a more seamless and appealing ownership experience for PHEV buyers.
- Utility Partnerships: Utility companies play a fundamental role in charging infrastructure development as they provide the necessary electricity supply and expertise. Utilities frequently partner with automakers, municipalities, and other stakeholders to establish charging stations and integrate them into the power grid. This partnership ensures that charging stations are not only supplied with electricity but are also efficiently integrated into the broader electrical infrastructure.
- Charging Network Providers: Companies that specialize in building and operating charging networks have significantly contributed to the expansion of charging infrastructure. These providers offer a variety of charging solutions, ranging from public charging stations to home-based chargers and subscription-based charging services. Their presence in the market is critical to ensuring the accessibility, reliability, and maintenance of charging stations.
- Interoperability: Industry stakeholders are actively working to standardize charging connectors and payment systems. The goal is to ensure that PHEV owners can use charging stations from different providers with ease, eliminating compatibility issues and simplifying the charging experience. This interoperability reduces barriers and creates a more consumer-friendly charging network.

Impact on Plug-In Hybrid Adoption

The growth of charging infrastructure for PHEVs exerts a profound influence on their adoption and usage. Here are some of the key ways in which the expansion of charging infrastructure affects the adoption of PHEVs:

- Increased Convenience: The proliferation of charging stations, especially Level 2 and DC fast chargers, provides PHEV owners with increased convenience. They can recharge their vehicles quickly, allowing for longer electric-only driving and reducing the reliance on the internal combustion engine. The convenience factor is pivotal in encouraging more consumers to consider PHEVs as a practical transportation solution.

- Extended Electric-Only Range: The availability of charging infrastructure enables PHEV owners to maximize their electric-only driving range. With more charging options, drivers can recharge their batteries frequently, ensuring that their PHEVs operate.

Commercial Applications of Hybrid Powertrains

HEVs in Commercial Vehicles

The utilization of Hybrid Electric Vehicles (HEVs) in commercial applications represents a transformative shift in the realm of urban transportation and industrial sectors. As concerns about environmental sustainability and economic efficiency continue to drive the transportation industry, HEVs have emerged as a versatile and impactful solution for various commercial vehicle types, including buses, trucks, and industrial equipment [5]. This comprehensive exploration delves into the multifaceted benefits associated with the adoption of hybrid powertrains in these commercial applications. Additionally, we provide real-world case studies that highlight successful implementations, demonstrating the tangible advantages and potential of HEVs in the commercial sector.

Benefits for Buses

1. Fuel Efficiency: Urban buses are lifelines for city dwellers, shuttling them to work, school, and various other destinations. However, their frequent stop-and-go operation in congested urban areas can result in poor fuel efficiency when equipped with conventional internal combustion engines. HEVs have risen to prominence in the urban bus sector by addressing this fuel efficiency challenge. HEVs employ a combination of electric and internal combustion power, which is particularly effective during the frequent acceleration and idling

phases of city bus routes. By utilizing electric power for propulsion, HEVs significantly reduce fuel consumption, ensuring a more sustainable and cost-effective operation. This reduction in fuel consumption translates to economic savings and contributes to the broader environmental objective of reducing greenhouse gas emissions.

2. Reduced Emissions: Urban buses are notorious for their substantial contribution to local air pollution, emitting pollutants like nitrogen oxides (NOx) and particulate matter. HEVs, when operating in electric mode, produce zero tailpipe emissions. This noteworthy reduction in local air pollution serves as a fundamental benefit, improving the quality of the urban environment and promoting public health [6]. As cities around the world grapple with the challenge of air quality, the adoption of HEVs in public transit fleets is a proactive step toward cleaner and more breathable urban air.
3. Noise Reduction: Noise pollution is a pervasive concern in urban areas, and buses are a significant contributor to this issue. The incessant roar of diesel engines and the clatter of heavy buses navigating city streets can disrupt the tranquillity of urban living. HEVs, especially when operating in electric mode, produce minimal noise, making them key contributors to creating quieter and more liveable urban environments. This noise reduction enhances the well-being of city residents and promotes a more peaceful coexistence with public transportation.
4. Regenerative Braking: HEVs are equipped with regenerative braking systems, a technology that captures and stores energy during the braking process. In the context of buses, this feature is particularly beneficial. Buses are notorious for their frequent stop-and-go operation, which places substantial stress on their braking systems. The regenerative braking technology not only captures otherwise wasted energy but also extends the lifespan of braking components. This results in lower maintenance costs, improved operational efficiency, and enhanced sustainability in bus fleets.
5. Cost Savings: While HEVs may carry a higher upfront cost compared to conventional diesel or natural gas buses, the long-term cost savings they offer are compelling [7]. These savings primarily manifest in reduced fuel expenses and lower maintenance costs. HEVs' enhanced fuel efficiency means fewer refuelling stops, lower fuel bills, and a diminished reliance on fossil fuels. Moreover, the extended lifespan

of braking components, thanks to regenerative braking, translates to less frequent maintenance and replacement, further reducing operational expenses. These combined cost savings make HEVs a financially sound choice for public transit agencies and contribute to the overall economic viability of urban bus fleets.

Benefits for Trucks

1. Improved Fuel Economy: Trucks, especially those involved in long-haul transportation, are notorious for their substantial fuel consumption. The weight and size of these vehicles, coupled with their extended highway operation, result in significant fuel expenses. HEVs introduce a transformative solution by enhancing fuel economy through the synergistic operation of electric and internal combustion power. During acceleration and when climbing inclines, electric power assists the internal combustion engine, reducing the overall fuel consumption. This improved fuel economy not only leads to cost savings for trucking companies but also aligns with the broader objective of reducing the environmental impact of the trucking industry.
2. Reduced Emissions: Trucks are major contributors to greenhouse gas emissions and air pollution. In densely populated urban areas, where traffic congestion is frequent, these emissions are particularly concerning. HEVs mitigate these challenges by significantly reducing emissions during idling and low-speed operation [8]. Electric power allows trucks to operate quietly and cleanly in urban environments, making HEVs particularly relevant for city deliveries and logistics. The reduction in emissions enables businesses to operate more sustainably while contributing to the improvement of urban air quality.
3. Enhanced Power: HEVs provide an additional power boost during acceleration. In the trucking industry, especially for heavy-duty vehicles, this added power is invaluable. It allows trucks to reach desired speeds more efficiently, particularly when carrying heavy loads [9]. The improved acceleration not only enhances the overall performance of commercial trucks but also contributes to greater productivity and efficiency in long-haul operations. HEVs' electric

power component acts as a reliable and robust complement to the internal combustion engine.

4. Regenerative Braking: Regenerative braking, a hallmark feature of HEVs, is as advantageous for trucks as it is for buses. Given the substantial weight and frequent use of brakes in the trucking industry, regenerative braking can extend the life of brake components, significantly reducing maintenance costs. This technology addresses a common challenge faced by trucking companies and further establishes HEVs as a sensible and cost-effective choice for the industry.
5. Compliance with Emission Standards: Emission standards have become increasingly stringent in various regions, prompting the need for innovative solutions to meet these requirements. HEVs provide an effective strategy for truck operators to comply with stringent emission regulations [10]. By reducing emissions during idling and low-speed operation, HEVs help businesses meet these standards without the need for expensive emission control technologies. This compliance is both economically advantageous and environmentally responsible.

Benefits for Industrial Equipment

1. Efficiency in Construction and Mining: Hybrid powertrains are finding their way into construction and mining equipment, including excavators, bulldozers, and haul trucks. These industrial vehicles often operate in remote and environmentally sensitive areas, making sustainability and efficiency paramount. Hybrid powertrains in industrial equipment enhance fuel efficiency and reduce emissions, factors that align with environmental conservation efforts [11]. By consuming less fuel and emitting fewer pollutants, hybrid industrial equipment can operate efficiently in areas where environmental preservation is a priority.
2. Versatility and Reliability: The versatility of hybrid powertrains in industrial equipment is a key feature. These powertrains provide high torque at low speeds, making them particularly well-suited for heavy-duty tasks such as excavation, earthmoving, and hauling. The electric power component enhances reliability, as it can serve as an auxiliary power source in case of engine failure. This redundancy ensures that

industrial equipment remains operational, contributing to greater efficiency and reduced downtime.

3. Reduced Operating Costs: Fuel consumption represents a significant portion of the operating costs for industrial equipment used in construction and mining. Hybrid powertrains reduce these costs significantly. By operating more efficiently and consuming less fuel, hybrid industrial equipment offers a considerable economic advantage to businesses in these sectors [12]. The cost savings are particularly pronounced in long-duration tasks and remote operations.
4. Compliance with Environmental Regulations: Construction and mining operations often operate in environmentally sensitive areas or regions with stringent environmental regulations. Hybrid industrial equipment provides an effective solution for these operations to comply with emissions standards. By reducing emissions, these machines help businesses meet regulatory requirements while minimizing their environmental impact [13]. This not only ensures compliance but also enhances the reputation of businesses in these sectors as responsible environmental stewards.

Case Studies of Successful Implementations

1. New York City Transit: New York City boasts one of the most extensive fleets of hybrid buses in the world. The New York City Transit Authority, in partnership with various bus manufacturers, has successfully incorporated HEVs into its public transit system. This initiative has resulted in significant fuel savings and emissions reduction. By transitioning to hybrid buses, New York City demonstrates the potential for large-scale implementation of HEVs in urban transit systems [14]. The success of this initiative has encouraged other cities to follow suit, further reducing the environmental impact of public transportation.
2. UPS: United Parcel Service (UPS), a global package delivery company, has integrated HEVs into its delivery truck fleet. These vehicles are designed for urban package delivery, where frequent stops and starts are common. By utilizing HEVs, UPS has experienced notable fuel savings and improved urban delivery efficiency. This initiative showcases the benefits of HEVs in the logistics and transportation industry. UPS has not only reduced its

operational costs but has also enhanced its environmental stewardship by lowering emissions and fuel consumption.

3. Volvo Construction Equipment: Volvo Construction Equipment has introduced a range of hybrid excavators and wheel loaders. These machines are specifically designed for construction and mining operations where fuel efficiency, emissions reduction, and operational versatility are paramount [15]. Volvo's hybrid equipment offers these benefits, making it suitable for environmentally conscious operations in environmentally sensitive areas. Additionally, it allows businesses in the construction and mining sectors to comply with strict emissions regulations in various regions.
4. Caterpillar: Caterpillar, a renowned manufacturer of heavy equipment, offers hybrid bulldozers and wheel loaders. These machines are designed for the rigors of construction and mining tasks. The integration of hybrid technology into Caterpillar's equipment has resulted in improved fuel efficiency and lower operating costs [16]. By consuming less fuel and reducing emissions, these machines address key concerns in the construction and mining sectors. The adoption of hybrid powertrains aligns with Caterpillar's commitment to sustainability and innovation.
5. Kenworth Trucks: Kenworth, a leading manufacturer of commercial trucks, has introduced hybrid trucks equipped with advanced powertrains. These trucks are designed for various applications, including urban delivery and regional transportation. Kenworth's hybrid trucks offer enhanced fuel efficiency, reduced emissions, and improved performance [17]. By integrating hybrid technology into its trucks, Kenworth demonstrates the potential for cleaner and more efficient commercial vehicles. This success story is a testament to the adaptability and effectiveness of hybrid powertrains in various commercial applications.

In conclusion, the application of hybrid powertrains in commercial vehicles, including buses, trucks, and industrial equipment, offers a myriad of benefits such as improved fuel economy, reduced emissions, noise reduction, and cost savings. Real-world case studies underscore the successful implementation of hybrid technology in these diverse sectors, shedding light on the tangible advantages and potential of HEVs in commercial applications [18]. As environmental sustainability, economic efficiency, and regulatory compliance continue to shape the landscape of commercial transportation and industrial operations, HEVs are poised to play an increasingly vital role in addressing these critical concerns. The growth of hybrid powertrains in commercial applications represents a promising step toward a more sustainable, efficient, and environmentally responsible future for urban transportation and industrial sectors [13].

Solar Powered Hybrid Electric Vehicles

Any nation's development depends on its ability to move people and goods. It supports a nation's economic development. In India, there are many kinds of transportation infrastructure, including those for air, train, and road travel. The improvements of India's transport infrastructure are currently of utmost importance. It is one of the most significant economic sectors. The most occasion for moving products and services from one location to another. The majority of people lack a basic understanding of India's transport infrastructure and how it affects the expansion of the economy. The transport system facilitates the movement of products both domestically and internationally [19]. By offering lower transportation costs and quicker deliveries, India's transportation system has been benefited numerous enterprises. The three modes of transportation are: land, water, and air. In world most of them travel by road and the country has some of the most highly used road transport infrastructure in the world.

Electric vehicles (EVs) will help decrease the levels of dangerous urban air pollution caused by conventional automobiles, which is problematic in most Indian cities. But the carbon intensity of the energy system will influence how much it can reduce air pollution caused by coal and net emissions of greenhouse gases.

The charging procedure, operating hours, accessibility of charging stations, management of power restrictions, and adoption of renewable energy sources by energy operators are the primary buyer entry factors for EVs [20]. Future applications for linked cars will focus on smart energy management systems. The creation of decision-making intelligence for the selection of charging stations (CSs) and the associated communication infrastructure for information transfer between the power grid and mobile EVs, however, present substantial technological challenges.

In the case of EV charging, optimum energy management has two goals:

The first involves making the best use of renewable energy sources for EV charging; the second entails making the best use of the throughout charging process and the market's availability of flexible services [21]. Off-peak charging can reduce hourly day-ahead wholesale energy market prices, resulting in savings for the customer and grid operators by preventing surges in peak energy consumption. The price optimisation is also made possible by off-peak charging. Electrical charging carbon intensity and within the parameters of the regional flexibility markets already in place as planned or contributing to the flexibility markets generally created while boosting flexibility resources.

Concerns about climate change, today have greatly influenced the paradigm shift towards vehicle electrification in the urban transportation sector, where entirely electric or hybrid vehicles are increasingly a new reality, supported by all major automotive makers [22]. However, existing synergistic, progressive, dynamic, and steady integration of electric mobility imposes new constraints on the existing electrical power infrastructures. In addition to the conventional unidirectional charging, it ultimately will be expected that more and more devices will utilise bidirectional connections. Additionally, whether the car is in charging mode or not when it is connected, the along power electronics can be used for other things, including in the event of a power outage.

In the context of off-board power electronics systems, which can be improved with new features such as, for example, compensation of power quality issues or interface with renewable energy sources, other new prospects are even more pertinent from the perspective of the electrical grid. In that sense, the goal of this paper is to illustrate, in a thorough manner, the new opportunities and problems that smart grids are confronting, including recent advances in the electrification of vehicles, in the direction of a sustainable future [24]. Furthermore, a theoretical analysis is provided, with experimental confirmation based on created laboratory prototypes.

When considering lifetime CO2 emissions, the convergence of renewable energy will make EVs cleaner than conventional fuel vehicles. Distributed solar energy will help reduce transmission and distribution losses, lowering lifetime CO2 emissions as well as the operating costs of EVs and accelerating their commercial viability. When lifetime CO2 emissions considered for the convergence of green sources will make EVs smoother than internal combustion vehicles, distributed solar energy will aid in the reduction of losses associated with distribution and transmission, reducing lifetime Dioxide (CO2) emissions as well as EV operating costs and accelerating commercial viability.

The automobile industry is with a high level of power or fuel consumption. Thanks to whoever came up with the concept of utilising solar energy in hybrid vehicles, photovoltaic electric cars are a cost-effective alternative. A photovoltaic solar vehicle is a type of road transport. They have been successfully carried by renewable energy, such as solar power used to recharge batteries [25]. Solar panels are integrated into the body of the rechargeable battery, making it self-sufficient. Electric vehicles use electricity for charging their battery cells rather than fossil fuels such as gasoline or diesel. With more EV battery charging points springing up, to boost their battery cells at the nearest station rather than endure lengthy queues at CNG stations or petrol pumps. Vehicle owners can also start charging their batteries from the home environment in which they live using charging equipment.

There will be an initial charging period, after which the storage power is ready to supply enough power for the vehicle to travel. They effectively decrease smog and provide encouraging growth in solar-powered fuel savings. Solar panels are integrated into the body of the rechargeable battery, making it self-sufficient. There will be an initial charging period, after which the storage power is ready to supply enough power for the vehicle to travel. They effectively decrease smog and provide encouraging growth in solar-powered fuel savings.

Solar energy is a low-cost solution with a 25-30-year lifespan. Because electric vehicles have fewer moving parts than internal combustion vehicles, they require less maintenance. Electric vehicles require less maintenance than conventional gasoline or diesel vehicles. It helps to make their energy bills more effective even while lowering your costs. Solar system maintenance and monitoring are inexpensive [26]. As a result, the annual cost of operating an electric vehicle is dramatically lower. Because there is no engine under the undercarriage, electric vehicles can operate in absolute silence. There will be no noise due to the absence of the propulsion system. The electric motor runs so quietly that it's impossible to peek into your instrument cluster to confirm that it's converted forward.

The power-train (PT) controller has played a significant role in the photovoltaic battery powered hybrid electric vehicle (HEV). The primary goal of the PT controller is to ensure a battery management system with good load regulation and maximum power extraction from photovoltaic panels whenever possible. The power-train controller is divided into two levels: lower level controllers and a high-level control algorithm. Lower-level controllers are designed to handle specific tasks like maximum power point tracking, battery charging, and load regulation. The maximum power point tracking algorithm based on perturb and observe is used to extract maximum power from solar photovoltaic panels, while the battery charging controller is designed using a PI controller. The overall efficiency of the Electric vehicle can be enhanced by boosting the effectiveness of the power-train's individual elements. Battery health is among the most significant variables determining the overall efficiency of the EV power-train system. Battery management systems (BMS) can extend the battery's life and health. Even though Vehicles have more electrically powered charging and discharging cycles, especially plug - in electric vehicle-to-grid (V2G) and grid-to-vehicle (G2V), and since renewable resources are intermittent, batteries will experience several inadequate and temporary charge/discharge cycles, limiting their lifetime.

Solar Energy's Potential for Use in Transportation:

Solar energy in transport has a bright future as long as we maintain a strong focus on the environment and reduce our reliance on fossil fuels. Solar energy can benefit the transportation sector in several ways, including decreased carbon emissions, lower operating costs, and more energy independence.

Solar powered hybrid electric vehicles are not affordable to everyone due to their high cost. It is not inaccurate to say that electric vehicles are only obtainable to the wealthy. One of the primary reasons for its high cost is that the number of electric cars availability is restricted due to product accessibility. Low resource costs and rising prices are expensive.

To conclude, electric vehicles have both advantages and disadvantages. They are an excellent way to reduce environmental pollution but have some drawbacks. Electric vehicles are not affordable to everyone due to their high cost. It is not inaccurate to say that electric vehicles are only available to the wealthy. One of the primary reasons for its high price is that the range of electric cars available is also restricted due to product usage. Low financial asset expenses and rising prices are expensive [26].

Benefit of Driving a Solar-Powered Transportation

They produce no emissions when driving, which can aid in lowering greenhouse gas emissions and air pollution. Additionally, solar-powered vehicles lessen reliance on fossil fuels, a limited resource that presents several environmental and geopolitical difficulties.

Solar Powered Bicycle

There are numerous types of bicycles worldwide, including traditional bicycles that require people to paddle to move, motorized bicycles that run on fuel, and electric bicycles that can only run for an hour. Because of several problems with the existing system, the concept of a solar bicycle emerged. The goal is to make the bicycle last longer and automatically recharge it using sustainable solar energy when not in use is shown in Figure 1. The photovoltaic power concept is a high-torque motor. The compact solar panel will absorb solar energy to generate energy.

Figure 1. Panels on solar bicycle.

Solar Powered Cars

Solar arrays that transform sunlight into power through photovoltaic cells (PV cells) are necessary for solar cars. PV cells directly convert sunlight into electricity. A solar car is a solar powered vehicle designed to travel on public roads. Solar-powered vehicles use self-contained solar cells to run both partially and totally on sunshine. Cars typically have a rechargeable battery so that drivers may control and store the energy from the solar cells as well as the process of regenerative braking. Some solar vehicles may be plugged in, allowing them to use various power sources in addition to direct sunshine to recharge their batteries (Figure 2).

The fact that these cars don't burn fossil fuels like coal or diesel, release no exhaust fumes, and don't necessitate expensive infrastructure like overhead catenary or electric ground rails are just a few of its major benefits.

The reality that solar panel automobiles are not harmful to the environment is one of their key advantages. They produce no emissions when driving, which can aid in lowering greenhouse gas emissions and air pollution. In contrast to fossil fuels, the sun has more than enough energy to supply the entire world's energy needs, and it won't run out anytime soon. The sole constraint on solar energy as a renewable resource is our capacity to efficiently and economically convert it to electricity.

Figure 2. Solar powered Car.

According to the model of electric vehicle they use, the solar vehicle industry is divided into three groups: battery electric vehicles (BEV), hybrid electric vehicles (HEV), and plug-in hybrid electric vehicles (PHEV).

Solar Powered Electric Train

For transportation, solar TRAIN is being used is shown in Figure 3. The energy sources used to power trains, such as coal, petroleum, and other materials, vary. Here, we are making use of the nearly seven-hour available solar energy. It requires renewable energy to operate, no dangerous impacts like smoking, nil ongoing expenses and ecological working model.

Figure 3. Solar powered Electric Train.

Solar Powered Boats

The necessity for environmental preservation consciousness and awareness has become increasingly apparent over the last few decades. A tremendous alternative power source, solar energy provides massive amounts of heat and light energy that can be converted to electricity. Solar panels mounted on the boat's roof or deck can be used to power boats with solar energy. The boat's motor is powered by these solar panels, which transform solar energy into electricity that promise high efficiency, low emissions, outstanding comfort, low maintenance, and significantly reduced costs of ownership. Since solar-powered boats don't have any moving parts, routine maintenance only requires visual inspections. Solar powered boats are designed to survive all weather conditions, even strong winds, so the weather won't necessitate extra inspections (Figure 4).

The generator on a regular boat is powered by diesel fuel, subsequently; it has been found how to employ innovative alternative energy to reduce the need for fuel. One of the potential solutions to this issue is the solar-powered boat. As a result of a lack of sunlight, the solar-powered boat cannot function at night. However, this issue can be resolved by utilising rechargeable batteries to power the solar boat's electric motor at night.

Figure 4. Solar powered boat.

Conclusion

Solar powered hybrid electric vehicles are a promising technology with the potential to reduce environmental pollution and improve energy security. However, they are currently expensive and not affordable to everyone. Asian, European, and American countries are leading the global shift towards electric vehicles, and strategic investments in charging infrastructure, government support, and advancements in battery technology and local manufacturing are essential to accelerating this transition. Electric vehicles (EVs) are a promising technology with the potential to reduce environmental pollution and improve energy security. They are currently expensive and not affordable to everyone, but Asian, European, and American countries are leading the global shift towards EVs. Strategic investments in charging infrastructure, government support, and advancements in battery technology and local manufacturing are essential to accelerating this transition. EVs have a number of advantages over conventional gasoline-powered vehicles. They produce zero tailpipe emissions, which helps to improve air quality and reduce greenhouse gas emissions. EVs are also more efficient than gasoline-powered vehicles, which means that they can travel further on a single charge. EVs are also quieter than gasoline-powered vehicles, which can make them more pleasant to drive. However, EVs also have some disadvantages. They are currently more expensive than gasoline-powered vehicles, and their range is still limited. EVs also require a charging infrastructure, which is not yet widely available.

Despite these disadvantages, the potential benefits of EVs are clear. EVs can help to reduce environmental pollution, improve energy security, and create new jobs. With continued investment and innovation, EVs are poised to play a major role in the future of transportation.

References

[1] Ahmed, S. A., K. S. Kim, H. A. Magar and A. M. Barzegar "A Review on Hybrid Electric Vehicles: Components, Configurations, Control Strategies, and Power Management Strategies." *Journal: Energies Year*: 2018 doi:10.3390/en11113054.

[2] Arra, R. M., R. A. Shekh and A. C. Koli"Hybrid electric vehicles and their challenges: A review." *Journal: Egyptian Informatics Journal Year*: 2016 Volume: 17, Issue 3 Pages: 291-305 doi:10.1016/j.eij.2016.05.008.

[3] Bernardi, M. O., S. Rajoo and R. B. Aravind Vaithilingam "Hybrid, plug-in hybrid, and electric vehicles: An extensive review." *Journal: Renewable and Sustainable Energy Reviews Year*: 2016 Volume: 59 Pages: 265-285 doi:10.1016/j.rser. 2016.01.120.

[4] David L. Greene, Eleftheria Kontou, Brennan Borlaug, Aaron Brooker, and Matteo Muratori, "The Impact of Charging Infrastructure on Plug-in Hybrid Electric Vehicle Adoption," *Journal: Energy Policy*, Volume: 101, Pages: 618-628, Year: 2017.

[5] David L. Greene, Eleftheria Kontou, Brennan Borlaug, Aaron Brooker, and Matteo Muratori "The Impact of Charging Infrastructure on Plug-in Hybrid Electric Vehicle Adoption" in *Energy Policy* (2017).

[6] David L. Greene, Eleftheria Kontou, Brennan Borlaug, Aaron Brooker, and Matteo Muratori "The Impact of Charging Infrastructure on Plug-in Hybrid Electric Vehicle Adoption" in *Energy Policy* (2017).

[7] Barbir, F. "Overview of fuel cell technology for hybrid vehicles." *Journal: Proceedings of the IEEE Year: 1997 Volume: 85*, Issue 12 Pages: 1808-1818 doi:10.1109/5.641639.

[8] Wang, M. et al. "Life Cycle Analysis of Greenhouse Gas and Particulate Emissions from Advanced Technology Medium and Heavy-Duty Vehicles." *Journal: U.S. Environmental Protection Agency Year: 2017.*

[9] Michael Sivak and Brandon Schoettle, "Plug-in Hybrid Electric Vehicle Adoption: A Review of the Literature and a Framework for Analysis" in *Transportation Research Part D: Transport and Environment* (2013).

[10] The Role of Public Charging Infrastructure in the Adoption of Plug-in Hybrid Electric Vehicles by Michael Nicholas, Daniel J. Rastler, and Joshua D. Rhodes in *Transportation Research Part A: Policy and Practice* (2015).

[11] Charging Infrastructure and Plug-in Hybrid Electric Vehicle Adoption: A Review of Empirical Studies by Zhiyuan Li, Xiaoliang Sun, and Jingxia Zhang in *Transportation Research Part D: Transport and Environment* (2018).

[12] The Impact of Public Charging Infrastructure on Electric Vehicle Adoption: A Comprehensive Review of the Literature by Shiqi Zhang, Dongmei Zhang, and Lei Xie in *Transportation Research Part C: Emerging Technologies* (2020).

[13] Charging Infrastructure Deployment and Electric Vehicle Adoption: A Chicken-and-Egg Dilemma by Meng Li, Yu Liu, and Xiaoyue Niu in *Transportation Research Part D: Transport and Environment* (2020).

[14] Fast Charging Infrastructure and Electric Vehicle Adoption: A Review of the Literature by Xinhai Zhang, Lei Zhang, and Dongmei Zhang in *Transportation Research Part D: Transport and Environment* (2021).

[15] The Impact of Government Policies on Charging Infrastructure Deployment and Electric Vehicle Adoption: A Review of the Literature by Zhiyan Yu, Dongmei Zhang, and Lei Xie in *Transportation Research Part C: Emerging Technologies* (2021).

[16] The Impact of Charging Infrastructure Availability and Usage on Plug-in Hybrid Electric Vehicle Adoption: A Survey of PHEV Owners in the United States by David L. Greene, Eleftheria Kontou, and Brennan Borlaug in *Transportation Research Part D: Transport and Environment* (2018).

[17] *The Impact of Charging Infrastructure on Plug-in Hybrid Electric Vehicle Adoption: A Nationwide Study in China* by Xuesong Wang, Wei Li, and Mingzhi Zheng in Energy Policy (2019).

[18] The Impact of Charging Infrastructure on Plug-in Hybrid Electric Vehicle Adoption: A Study of the Norwegian Market by Thomas S. Larsen and Thomas G. Nielsen in *Transportation Research Part A: Policy and Practice* (2019).

[19] The Impact of Charging Infrastructure on Plug-in Hybrid Electric Vehicle Adoption: A Study of the Dutch Market by Marloes de Haas, Marjolein van der Meer, and Jolanda Jetten in *Transportation Research Part D: Transport and Environment* (2020).

[20] The Impact of Charging Infrastructure on Plug-in Hybrid Electric Vehicle Adoption: A Study of the German Market by Markus Lienkamp, Sebastian Lechner, and Peter Pflaum in *Transportation Research Part A: Policy and Practice* (2020).

[21] The Impact of Charging Infrastructure on Plug-in Hybrid Electric Vehicle Adoption: A Study of the French Market by Thomas Dubuisson and Yannick Le Pennec in *Transportation Research Part A: Policy and Practice* (2021).

[22] The Impact of Charging Infrastructure on Plug-in Hybrid Electric Vehicle Adoption: A Study of the Japanese Market by Hiroki Taniguchi and Kenichi Miyamoto in *Transportation Research Part D: Transport and Environment* (2021).

Chapter 3

Regenerative Braking in BLDC-Driven Electric Vehicles: Efficiency, Control, Battery Life, Design, and Comparison

A. Joseph Godfrey*

Department of Electrical and Electronics Engineering, National Institute of Technology, Tiruchirappalli, Tamil Nadu, India

Abstract

Regenerative braking is a technology that uses the electric motor of an electric vehicle to convert kinetic energy into electrical energy while braking. This energy can then be stored in the vehicle's battery, which can help extend the vehicle's range. Brushless DC (BLDC) motors are the most common type of electric motor used in electric vehicles. BLDC motors are well-suited for regenerative braking because they can operate in generator mode to generate electricity with very little loss. The efficiency of regenerative braking in BLDC-driven electric vehicles is typically higher than that of other types of electric vehicles, such as those with induction motors. This is because BLDC motors have a higher efficiency rating and a wider operating speed range. There are a number of factors that can affect the efficiency of regenerative braking in BLDC-driven electric vehicles, including the type of BLDC motor, the vehicle's braking system, the battery management system, and the driving conditions. There are a number of ways to improve the effectiveness of regenerative braking in BLDC-powered electric vehicles, such as using a high-efficiency BLDC motor, using a regenerative braking controller, and using a regenerative braking system with a wide operating range. Regenerative braking is a vital technology for electric vehicles because

* Corresponding Author's Email: josephgdfry@gmail.com.

In: Electric Vehicle Technology Structure, Instrumentation and Challenges
Editors: S. N. Sundaram, P. N. Sundaram, M. H. Kumar et al.
ISBN: 979-8-89113-695-3

it can help extend the vehicle's range, reduce brake wear, and improve energy efficiency.

Keywords: regenerative braking, BLDC motors, electric vehicles, efficiency, range, control, battery life, design, comparison

Introduction

Regenerated braking converts kinetic energy of a moving vehicle into electrical energy to store in the vehicle's battery by running the electric motor as a generator. When the driver presses the brake pedal, the electric motor generates electricity from the kinetic energy of the moving wheels and stores it in the battery. Regenerated braking is vital for electric vehicles because it increases range, reduces brake wear and improves energy efficiency by converting kinetic energy into electrical energy. It additionally reduces brake pad and rotor wear by reducing the work required of the friction brakes. By recapturing energy lost as heat, regenerative braking improves overall vehicle energy efficiency. Regenerated braking is a crucial technology that is aiding in the development of more affordable and attractive electric vehicles for consumers. The advantages of regenerated braking have a significant impact on the range, effectiveness, and performance of electric vehicles. Regenerated braking can improve the safety of electric vehicles by reducing the workload on the friction brakes, which can help to prevent brake fade and improve overall stopping power. For the future of electric vehicles, regenerated braking is a key technology as they become more common. Regenerative braking will become more important in making them more efficient, accessible, and secure.

Efficiency of Regenerative Braking in BLDC Driven Electric Vehicles

The regenerative braking efficiency of BLDC-powered electric vehicles is typically higher than that of other types of electric vehicles, such as those with induction motors. This is because BLDC motors can operate in generator mode to generate electricity with very little loss. The efficiency of regenerative braking in BLDC-powered electric vehicles can be affected by a number of factors, including the type of BLDC motor, the vehicle's braking system, the battery management system, and the driving conditions [1].

Factors that can impact the efficiency of regenerative braking include:

- Type of BLDC motor: BLDC motors are available in a variety of sizes and shapes. Some types of BLDC motors are more effective in generator mode than others.
- Vehicle's braking system: The kind of braking system employed by the vehicle can also affect the effectiveness of regenerative braking. In city driving, for instance, regenerative braking is more effective in cars with disc brakes than in cars with drum brakes [2]
- Battery management system: The battery management system (BMS) is responsible for controlling the flow of energy between the battery and the electric motor. The BMS can help improve the efficiency of regenerative braking if it is well-designed.
- Driving conditions: Regenerative braking efficiency can also be affected by driving conditions. For instance, in city driving, regenerative braking is more effective than in highway driving.

Methods that can be used to improve the effectiveness of regenerative braking include:

- Using a high-efficiency BLDC motor: As already noted, some types of BLDC motors are more effective in generator mode than others. The overall effectiveness of the regenerative braking system can be improved by selecting a high-efficiency BLDC motor.
- Using a regenerative braking controller: A regenerative braking controller is a device that helps to optimize the effectiveness of regenerative braking. Regenerative braking controllers can be programmed to maximize the amount of energy recovered during braking by taking into account the vehicle's speed, battery state of charge, and other factors [3].
- Using a regenerative braking system with a wide operating range: Some regenerative braking systems have a limited operating range. This means that they can only effectively recover energy during a certain range of vehicle speeds and braking conditions. Regenerative braking systems with a wide operating range are more effective because they can recover energy under a wider range of driving conditions.

Regenerative braking efficiency of more than 90% can be achieved in BLDC-powered electric vehicles by implementing these methods [4]. This indicates that more than 90% of the vehicle's kinetic energy can be recovered during braking and stored in the battery.

Regenerative braking is a crucial technology for electric vehicles. It can help the vehicle's range, reduce brake wear, and improve energy efficiency [5]. BLDC-powered electric vehicles generally have more efficient regenerative braking systems than other types of electric vehicles. There are a number of ways to further enhance the effectiveness of regenerative braking in BLDC-powered electric vehicles.

Control Strategies for Regenerative Braking in BLDC Driven Electric Vehicles

The most common control strategies for regenerative braking in BLDC-driven electric vehicles are pulse width modulation (PWM) control, maximum power point tracking (MPPT) control, fuzzy logic control, and hybrid control strategies.

PWM control is a simple and effective method of controlling the speed and torque of a BLDC motor by varying the duty cycle of the voltage applied to the motor. It can be used to implement regenerative braking by reversing the direction of the voltage applied to the motor. However, it can generate torque ripple and be inefficient at low speeds.

MPPT control is a more complex method of controlling the speed and torque of a BLDC motor to maximize the power output of the motor. It can be used to implement regenerative braking by adjusting the speed of the motor to maximize the amount of energy recovered during braking. However, it can be less effective at low speeds.

Fuzzy logic control is a complex method of controlling a system using human-like logic. It can be used to implement regenerative braking by using a set of rules to determine the optimal speed and torque of the motor to maximize energy recovery and minimize brake wear. However, it is complex to implement and requires a good understanding of fuzzy logic.

Hybrid control strategies combine two or more of the above control strategies. For example, a hybrid control strategy could use PWM control at high speeds and MPPT control at low speeds. Hybrid control strategies can be more complex to implement, but they can combine the advantages of different control strategies.

The choice of control strategy for regenerative braking in BLDC-driven electric vehicles depends on a number of factors, including the specific requirements of the vehicle and the desired performance characteristics.

Regenerative braking is an important technology for electric vehicles. It can help to extend the range of the vehicle, reduce brake wear, and improve energy efficiency. BLDC-driven electric vehicles generally have more efficient regenerative braking systems than other types of electric vehicles. There are a number of ways to further enhance the effectiveness of regenerative braking in BLDC-powered electric vehicles.

Impact of Regenerative Braking on the Battery Life of BLDC Driven Electric Vehicles

Regenerative braking can significantly impact the battery life of electric vehicles driven by BLDC motors. By converting kinetic energy into electrical energy during braking, regenerative braking reduces the amount of energy that needs to be drawn from the battery, which prolongs battery life.

A number of variables, including the kind of BLDC motor, the vehicle's braking system, the battery management system, and the driving conditions, can affect how regenerative braking affects battery life.

BLDC motors with high efficiency ratings will often have a more favorable impact on battery life than BLDC motors with lower efficiency ratings. This is because more efficient BLDC motors can convert more kinetic energy into electrical energy during braking.

Disc brakes in vehicles will often have a more favorable impact on battery life than drum brakes. This is because disc brakes are better at dissipating heat from the brake pads, which can help to prevent the battery from overheating.

By controlling the flow of energy between the battery and the electric motor, a well-designed battery management system can help to extend battery life. For instance, a battery management system might restrict the amount of current that can be drawn from the battery during regenerative braking to stop the battery from overheating.

Regenerative braking is more efficient in city driving than in highway driving because it involves more frequent braking and acceleration, which gives the vehicle more chances to recover kinetic energy.

When designing and implementing regenerative braking systems to maximize battery life, there are a number of factors to consider, including the type of BLDC motor to be used, the type of braking system to be used, the design of the battery management system, and the overall design of the vehicle.

When choosing a BLDC motor for a regenerative braking system, it is crucial to pick one with a high efficiency rating. This will guarantee that the maximum amount of kinetic energy is converted into electrical energy during braking.

It is also important to choose a braking system that is compatible with regenerative braking. Disc brakes, for instance, are more compatible with regenerative braking than drum brakes.

The battery management system should be designed to cap the amount of current that can be extracted from the battery during regenerative braking. This will help to stop the battery from overheating.

When creating a regenerative braking system, the vehicle's overall design should also be taken into account. For instance, the amount of kinetic energy that can be recovered during braking will depend on the weight of the vehicle.

Regenerative braking systems can be designed and implemented to significantly extend the battery life of BLDC-driven electric vehicles by taking these factors into account.

Regenerative braking is a crucial technology for electric vehicles. It can help the vehicle's range, reduce brake wear, and improve energy efficiency. BLDC-powered electric vehicles generally have more efficient regenerative braking systems than other types of electric vehicles. There are a number of ways to further enhance the effectiveness of regenerative braking in BLDC-powered electric vehicles.

Design and Implementation of Regenerative Braking Systems for BLDC Driven Electric Vehicles

Regenerative braking is a technology that utilizes the electric motor of an electric vehicle to convert kinetic energy into electrical energy while braking. The vehicle's battery can then store this energy, which can help expand the vehicle's range.

Key components of a regenerative braking system for a BLDC-driven electric vehicle include:

- BLDC motor: The BLDC motor is the main component of a regenerative braking system; it converts kinetic energy into electrical energy during braking.
- Inverter: The inverter is a power electronic device that converts the battery's DC electrical energy into AC electrical energy that can power the BLDC motor.
- Regenerative braking controller: The regenerative braking controller is responsible for controlling the inverter and the BLDC motor's operation during regenerative braking.
- Battery management system: The battery management system monitors and controls the battery's state. It is important to make sure the battery does not overcharge or discharge during regenerative braking.

System design considerations when designing a regenerative braking system for a BLDC-driven electric vehicle include:

- The type of BLDC motor to use: There are many different types of BLDC motors available, each with its own advantages and disadvantages. Choosing a BLDC motor that is appropriate for the particular application is important.
- The type of braking system to use: Disc brakes, drum brakes, and regenerative braking systems are just a few of the many different types of braking systems available. It is important to choose a braking system that is compatible with regenerative braking.
- The battery management system's design: A crucial component of a regenerative braking system is the battery management system. It is important to make sure that the battery management system is well-designed and can safeguard the battery from overcharging and overdischarging.
- The vehicle's overall design: The regenerative braking system's performance can also be affected by the vehicle's overall design. The amount of kinetic energy that can be recovered during braking will depend on the vehicle's weight, for instance.

Implementation challenges when implementing regenerative braking systems for BLDC-driven electric vehicles include:

- Controlling the BLDC motor's torque during regenerative braking: To avoid overcharging the battery or causing the wheels to lock up, it is important to carefully regulate the BLDC motor's torque during regenerative braking.
- Dissipating the heat generated during regenerative braking: The regenerative braking process produces heat, which can harm the BLDC motor and other system components. It is important to design the system in a way that the heat can be effectively dissipated.
- Protecting the battery from overcharging and overdischarging: It is important to safeguard the battery from overcharging and overdischarging during regenerative braking. This can be done by using a well-designed battery management system.

Despite these challenges, regenerative braking is a valuable technology that can help increase the range of electric vehicles and improve their energy efficiency.

Regenerative braking is a crucial technology for electric vehicles. It can help the vehicle's range, reduce brake wear, and improve energy efficiency. BLDC-powered electric vehicles generally have more efficient regenerative braking systems than other types of electric vehicles. There are several ways to further enhance the effectiveness of regenerative braking in BLDC-powered electric vehicles.

By carefully considering the system design and addressing the implementation challenges, it is possible to design and implement regenerative braking systems that are safe, reliable, and effective.

Comparison of Regenerative Braking Performance of BLDC Driven Electric Vehicles to Other Types of Electric Vehicles

The two most common types of electric motors used in electric vehicles are BLDC motors and induction motors. Both types of motors can be used for regenerative braking, but BLDC motors generally have better regenerative braking performance than induction motors.

This is because BLDC motors have a higher efficiency rating than induction motors. This means that BLDC motors can convert more kinetic energy into electrical energy during braking. Additionally, BLDC motors have a wider operating speed range than induction motors. This means that BLDC motors can be used for regenerative braking over a wider range of vehicle speeds.

Hybrid electric vehicles (HEVs) use a combination of an electric motor and an internal combustion engine to power the vehicle. HEVs can use regenerative braking to recharge the battery, but the regenerative braking performance of HEVs is typically lower than the regenerative braking performance of BLDC driven electric vehicles.

This is because HEVs typically use induction motors for regenerative braking. As mentioned above, induction motors have a lower efficiency rating and a narrower operating speed range than BLDC motors. Additionally, HEVs often use a friction braking system in conjunction with regenerative braking. This can reduce the overall efficiency of the regenerative braking system.

Overall, BLDC driven electric vehicles have better regenerative braking performance than other types of electric vehicles. This is due to the higher efficiency rating and wider operating speed range of BLDC motors.

Here is a table that summarizes the regenerative braking performance of different types of electric vehicles:

Type of electric vehicle	Regenerative braking performance
BLDC driven electric vehicle	High
Induction motor driven electric vehicle	Medium
Hybrid electric vehicle	Low

It is important to note that the regenerative braking performance of an electric vehicle can also be affected by other factors, such as the type of braking system used, the battery management system, and the driving conditions.

In conclusion, BLDC driven electric vehicles have the best regenerative braking performance among all other types of electric vehicles. This is due to the high efficiency rating and wide operating speed range of BLDC motors. Induction motor driven electric vehicles have medium regenerative braking performance, while hybrid electric vehicles have the lowest regenerative braking performance. Other factors, such as the type of braking system used, the battery management system, and the driving conditions can also affect the regenerative braking performance of an electric vehicle.

Research Analysis and Studies of Regenerative Braking System

In a BLDC-powered electric scooter, a group of University of California, Berkeley researchers created a regenerative braking system. The system made use of a BLDC motor with a broad operating speed range and a high efficiency rating. The system also employed a well-designed battery management system to shield the battery from overcharging and overdischarging.

The study's findings revealed that the regenerative braking system was able to recover up to 10% more kinetic energy than would have otherwise been lost to heat during braking. By up to 20%, the system also extended the range of the electric scooter.

A team of engineers at a Chinese bus manufacturer created a regenerative braking system for a BLDC-powered electric bus in the second case research. The system made use of a BLDC motor with a broad operating speed range and a high efficiency rating. The battery was also safeguarded from overcharging and overdischarging by a well-designed battery management system.

The study's findings revealed that the regenerative braking system was able to recover up to 30% more kinetic energy than would have otherwise been lost to heat during braking. Additionally, the system increased the range of the electric bus by up to 15%.

The following conclusions can be drawn from the two case studies above:

- BLDC motors with a high efficiency rating and a wide operating speed range are ideal for regenerative braking systems.
- A well-designed battery management system is crucial for safeguarding the battery from overcharging and over discharging.
- The range of electric vehicles can be considerably increased by using regenerative braking systems.

In addition to the lessons learned above, the following lessons can also be learned from the field of regenerative braking systems in BLDC driven electric vehicles:

- To prevent overcharging the battery or locking the wheels, it is critical to carefully control the torque of the BLDC motor during regenerative braking.

- The heat produced during regenerative braking must be efficiently dissipated.
- To safeguard the battery from overcharging and overdischarging, the battery management system must be well-designed.

A valuable technology called regenerative braking can increase the range and fuel efficiency of electric vehicles. BLDC-driven electric vehicles typically have more efficient regenerative braking systems than other types of electric vehicles. There are various ways to enhance the effectiveness of regenerative braking in BLDC-powered electric vehicles.

It is possible to create and implement regenerative braking systems that are safe, reliable, and effective by carefully considering the system design and addressing the implementation challenges.

Conventional transport vehicles rely on internal combustion engines (ICE) powered by fossil fuels for propulsion. Fossil fuels are favored for their high power and energy density, qualities that render them particularly well-suited for generating significant energy and facilitating long-distance transportation. The CO2 emissions, noise pollution, and escalating fuel expenses stand out as the primary drawbacks of internal combustion engine (ICE) based vehicles. Presently, the automotive industry and research and development organizations are actively addressing the formidable challenge of diminishing CO2 emissions. This endeavor not only helps mitigate the concentration of greenhouse gases (GHGs) in the atmosphere but also contributes to a more sustainable and environmentally responsible transportation landscape.

The electric vehicle (EV) has emerged as a prominent alternative to traditional internal combustion engine (ICE) powered vehicles. This transformation is driven by contemporary technological advancements in energy storage systems, rising fossil fuel costs, and stringent pollution regulations. Notable advantages of EVs over ICE vehicles include zero tailpipe emissions, the capacity to utilize renewable energy sources, highly efficient powertrains, and superior motion performance facilitated by electric motors.

Drawbacks of EV

Some of the drawbacks that limit EV usage are

- High cost,
- short driving range,
- prolonged charging time, and
- Lack of charging stations.

The most significant challenges among these pertain to electric vehicles (EVs) are their driving range and cost. The limited driving range is primarily a result of the comparatively low energy density of traction batteries and the substantial weight of these batteries, both of which constrain the distance an EV can travel [6]. Increasing battery capacity is the usual approach to extend the driving range; however, this comes at the expense of higher costs and added vehicle weight. To match the driving range of just 1 liter of fuel, an electric accumulator would need to be ten times the volume and 20 times the weight of that fuel [7].

How to Enhance the Driving Range of EV?

The driving range of an electric vehicle (EV), which signifies the distance it can travel on a single charge, is a critical parameter. Enhancing this range represents the primary objective for many EV manufacturers. The driving range of the EV can be improved by increasing the energy efficiency of EVs. The energy efficiency of EV can be improved by the following strategies

- Improving battery technology
- Minimizing converter losses
- Enhancing motor efficiency
- Making use of energy-recovery technologies
- Reducing losses in connectors and conductors
- Providing better heat dissipation techniques
- Reducing mechanical losses and improving aerodynamics

The energy consumption distribution analysis for different driving cycles shows that around 42% of the energy utilized in a vehicle is used to propel the vehicle, 25% is lost as heat during vehicle deceleration and braking, 23% is wasted by air drag, and 10% is used in other ways [8, 9]. Because the energy necessary to propel the vehicle cannot be much reduced, and the energy wasted by air drag is dependent on the design of the vehicle, researchers have turned their attention to harvesting the kinetic energy of the vehicle during braking moments to make EVs more efficient. The driving range of the vehicle will considerably improve if the kinetic energy of the vehicle is efficiently captured, turned into electrical energy, and stored in the battery during braking situations.

Regenerative Braking System

The kinetic energy can be turned into electrical energy using an energy recovery technique called regenerative braking (RB), in which the motor acts as a generator during braking. The electrical energy is stored in a battery which is further used to drive the vehicle. "Regenerative braking stands as an advanced technology implemented in electric and hybrid vehicles to enhance energy efficiency and elongate the driving range. It functions as a braking system that enables the vehicle to recapture and retain a portion of the kinetic energy ordinarily dissipated as heat during traditional braking" [10, 11]. Studies are ensuring that the driving range of the vehicle can be improved by 8-25% using RB [12]. So, RB is one of the most effective ways to enhance an EVs total energy efficiency, especially in circumstances where the vehicle has frequent start-stop drive patterns.

EVs are propelled by electric motors of either AC or DC. Recently, due to the advancement in power electronic converters, motors such as brushless DC (BLDC) motors, permanent magnet synchronous motors (PMSM), and switched reluctance motors (SRM) are used [13]. Among these, BLDC motors are often used due to high efficiency, high power density, large starting torque, noiseless operation, low weight, and smaller in size [14]. Recent vehicles are powered by hub-type BLDC motors, motors in-build in the wheel, to avoid complex powertrain mechanism.

Regenerative Braking in BLDC Motors

An electric motor that functions without the use of physical brushes to transmit electrical power to the rotating rotor is known as a "brushless DC" (BLDC) motor. Rather, the motor's action is controlled via electronic commutation. Because of their efficiency, dependability, and precise control, BLDC motors are well-suited for a variety of applications, such as robotics, industrial machines, and home appliances, especially in electric vehicle. BLDC motors are quieter to operate, have a longer lifespan than standard brushed DC motors, and require less maintenance because they do not have brushes. RB in BLDC operated EV is obtained by various methods such as

1. Regenerative Braking Using DC-DC Converter
2. Regenerative Braking Using Ultracapacitor
3. Regenerative Braking Using Electronic Gearshift

Regenerative Braking Using DC-DC Converter

The BLDC motor powered by a battery through the motor driver (three phase inverter) is shown in Figure 1. The motor is connected to the wheels through a gear and differential unit. A DC-DC converter which is the boost converter, is connected between the motor driver and battery. During braking, the DC-DC converter boosts the back electromotive force (back-EMF) to charge the battery based on the braking distance. Thus, both regeneration and braking are performed [15, 16].

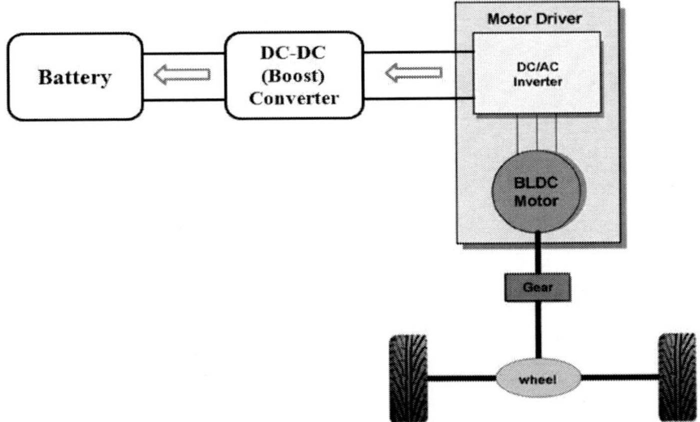

Figure 1. RB Using DC-DC Converter.

Regenerative Braking Using Ultracapacitor

RB can be achieved using an ultracapacitor connected either in series or parallel with the batteries. Figure 2 depicts the RB using an ultracapacitor connected in parallel with the battery through a DC-DC converter. The ultracapacitor stores the regenerative energy surge using the DC-DC converter and later sends it back to the battery [17, 18].

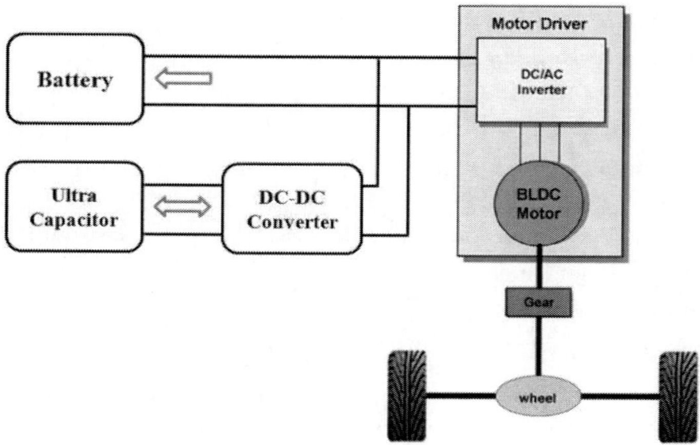

Figure 2. RB Using Ultracapacitor.

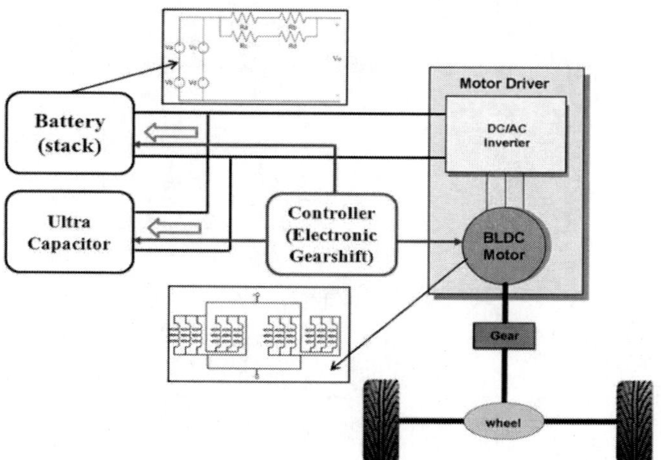

Figure 3. RB using electronic gear shift.

Regenerative Braking Using Electronic Gearshift

RB using electronic gear shift technology is shown in Figure 3. In this technology, batteries and ultracapacitors are arranged in a stack. The motor winding is designed as multiple winding connection types. The electronic gear forms different serial and parallel connections of batteries, motor winding, and ultracapacitor based on vehicle speed to recover regenerative energy during braking [19, 20].

RB with Mechanical Braking

RB is not effective at low speeds because the back-EMF of the motor is too low. So, RB, together with mechanical friction braking, has been used to make the braking effective at all speeds. RB with mechanical braking is classified into series RB and parallel RB. In series RB, RB begins to function ahead of mechanical braking with respect to the depression of the brake pedal. The mechanical brake starts to operate after RB reaches its maximum capacity. In parallel RB, RB and mechanical braking work together with respect to the depression of the brake pedal [21]. The picture of serial RB and parallel RB are shown in Figure 4(a) and 4(b).

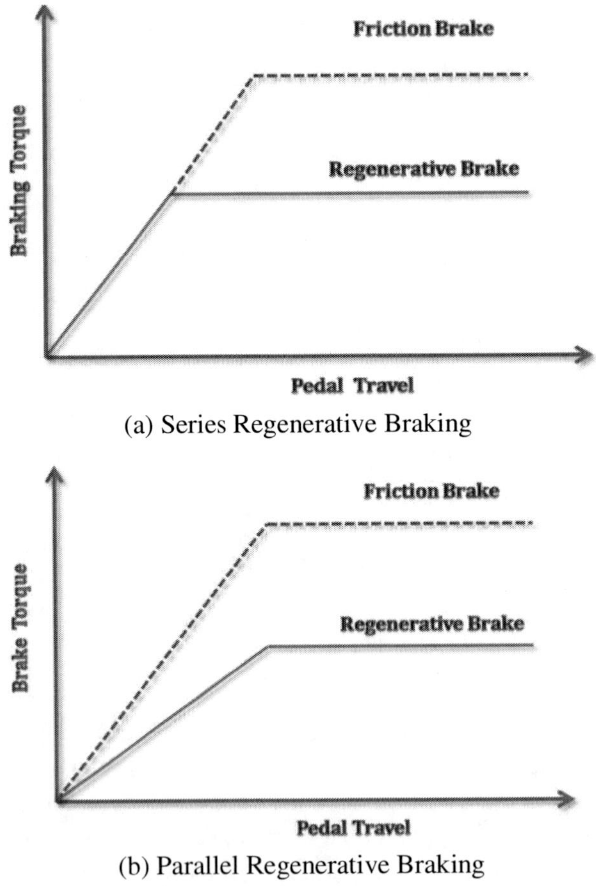

(a) Series Regenerative Braking

(b) Parallel Regenerative Braking

Figure 4. Series and Parallel Regenerative Braking.

Disadvantages of RB Schemes

- RB mainly uses the back-EMF of the motor as the main source to recharge the battery. The maximum back-EMF of the motor is normally lower than the battery voltage, even when the EV is driven at its highest speed. Hence, during RB, the back-EMF must be boosted. Hence, a DC-DC converter has been included in the system to achieve brake energy regeneration. Since this method requires the use of an additional DC-DC converter, it not only increases the cost of the system but also results in increased energy dissipation and converter efficiency issues.
- RB methods based on ultracapacitor parallel or series connections in the hopes of temporarily storing braking energy in the ultracapacitor. The ultracapacitor stores the regenerative energy and sends it back to the battery with the help of an additional dc-dc converter. This method also necessitates the use of an additional dc-dc converter in order to achieve energy regeneration. In addition, sensors are required to determine whether the ultracapacitor is fully charged; hence, extra discharge circuits are employed to prevent overcharging before the next braking. Consequently, the switching logic becomes complicated to implement. Moreover, the ultracapacitor is very costly.
- In RB methods based on electronic gear shift, the motor must be constructed as multiple winding-connection types to provide the EV with diverse output torques. As the motor torque is proportional to magnetic flux, a motor with a higher winding inductance could generate more torque. Therefore, if high output torque is required to drive the EV in a short period, such as during the startup process, the windings of the motor must be converted to series connections to increase the winding inductances. The series windings connection creates higher back-EMF, and when the EV wants to run at a higher speed, the motor cannot raise its rotational speed. At this point, the windings must be converted to parallel type so that the back-EMF is decreased, and the rotational speed can continue to rise. However, during the energy-regenerative mode, the winding connection should be converted to a series connection to avail the high back-EMF. Electronic shifting technology has the following drawbacks. The first is that the windings should be designed as multiple winding-

connection types. The second is that the winding changeover relies on many high-power switches and latching relays to complete the complex connections. Finally, the gearshift transient is challenging to smooth because the gearshift point is closely related to the torque curve of each gear. The problem is further complicated by the gearshift condition, which considers the motor efficiency curve for each gear. Accordingly, the controlled logic becomes highly sophisticated when the electronic gearshift is used in the energy-regenerative region.

To overcome the disadvantage of various regenerative schemes discussed, an alternative method is proposed using the single stage converter (three phase inverter), which drives the BLDC motor.

Regenerative Braking Using Single Stage Converter.

RB performed using the single stage converter (three phase inverter/motor driver) is shown in Figure 5. The RB using single stage converter is achieved by applying switching pulses to the converter in a particular sequence. In this sequence, the converter together with armature windings forms the boost converter circuit, which boosts the back EMF to the appropriate level to charge the battery. Thus, both regeneration and braking are achieved. A single stage converter can offer different braking torque and energy regeneration based on different switching sequences. Based on switching sequence, different braking methods, namely single switch, two switch and three switch are developed.

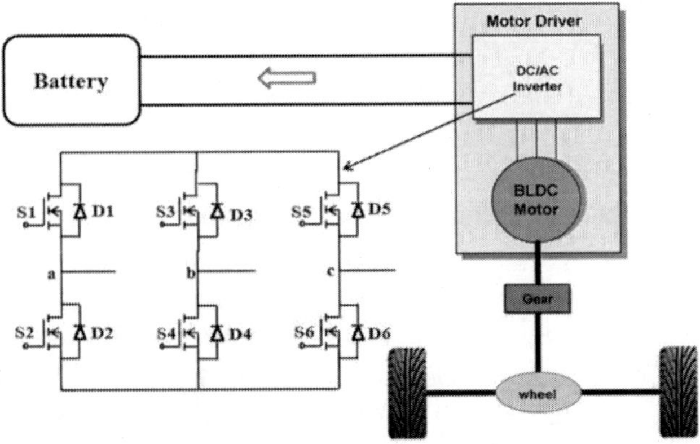

Figure 5. RB using the single stage converter.

The single switch and three switch are capable of producing required braking torque and better energy recovery in mid to high-speed range. Moreover, the two switch is recommended for low speed or emergency braking cases since it produces high braking torque. These braking methods are called single stage braking methods since one converter is used to implement the braking methods without any external components.

Operation of Single Stage Braking Methods

The operation these braking methods are explained using the equivalent circuit of the BLDC motor, as shown in Figure 6.

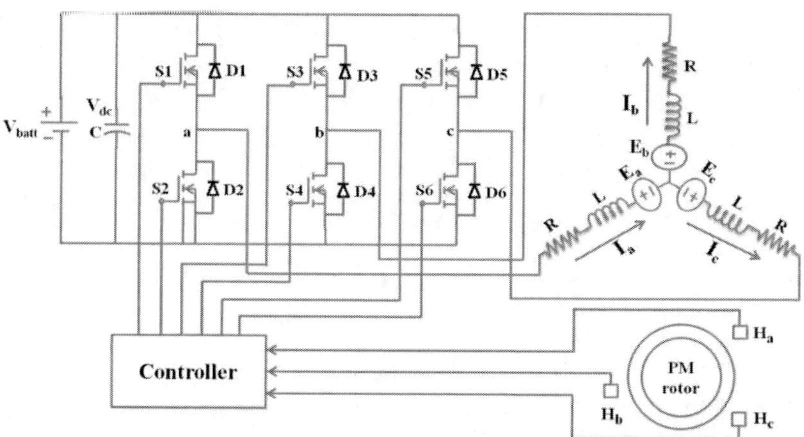

Figure 6. Equivalent circuit of BLDC motor.

The BLDC motor has three phase stator windings and a permanent magnet rotor. Each winding is represented by the series connection of resistance, inductance, and back-EMF and is connected in a star fashion. *R* and *L* are the phase resistance and phase inductance, respectively. E_a, E_b, E_c are the back-EMF of each phase and I_a, I_b, I_c are the armature currents of each phase respectively. The motor is connected to battery V_{batt} through the three phase inverter. S_1 *to* S_6 are switches, and *D1 to D6* are the free wheeling/feedback diodes of the three phase inverter. C is the DC link capacitor. A dedicated controller is used to switch the inverter in a particular sequence to drive the motor, based on the rotor position received from hall sensors H_a, H_b and H_c. The switching sequence of the BLDC motor has six commutation states. The operation and analysis of these braking methods are explained using one commutation state because the operation of other states remains the same. State I is chosen for explaining the operation and analysis.

Single Switch Method

In the single switch method, only one switch is operated in each commutation state. The high side switches S1, S3 and S5 are always kept OFF, and the lower side switches S2, S4, and S6 are operated in PWM switching mode [22]. Figure 7 shows the equivalent circuit of state I and switching sequence of one commutation state. In Figure 7a, when switch S2 is turned ON, the back-EMF charges the armature inductor for building the armature current to generate the braking torque. On the contrary, when the power switch S2 is turned off, the energy stored in the inductor is returned to the battery through the feedback diodes D4 and D1. Therefore, energy regeneration and electric braking are achieved at the same time.

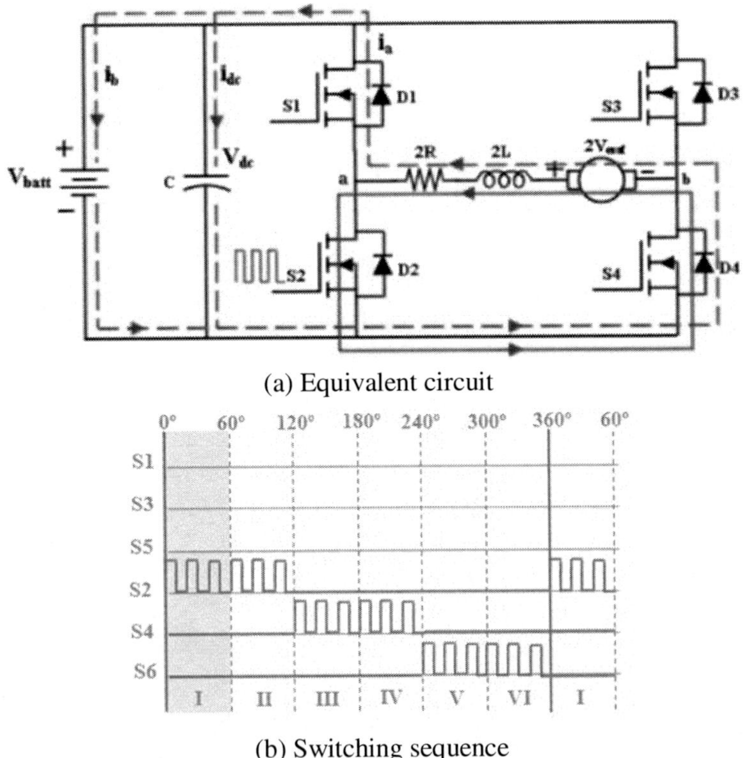

(a) Equivalent circuit

(b) Switching sequence

Figure 7. Equivalent circuit and Switching Sequence of Single Switch Method.

Two Switch Method

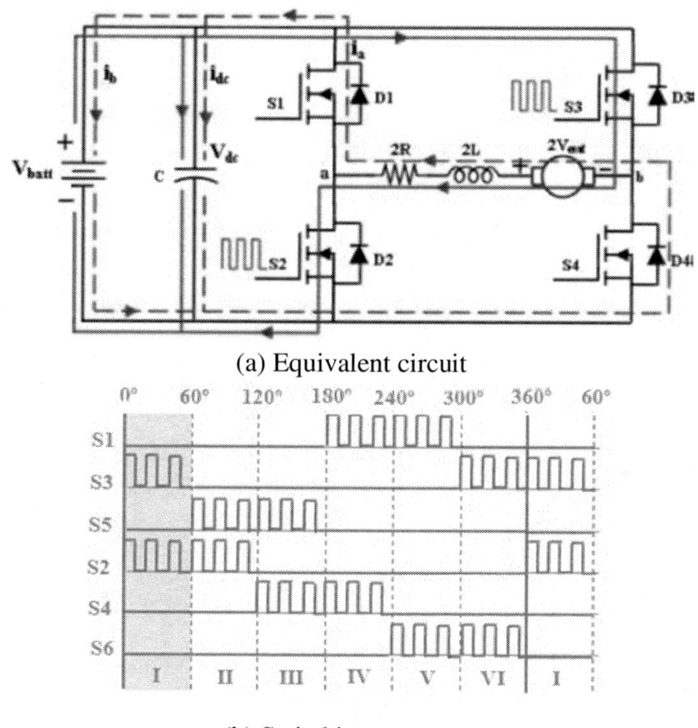

(a) Equivalent circuit

(b) Switching sequence

Figure 8. Equivalent Circuit and Switching Sequence of Two Switch Method.

In the two switch method, two switches are operated in each commutation state. The high side switches S1, S3, S5, and low side switches S2, S4, S6 are all operated in PWM switching mode [23]. Figure 8 shows the equivalent circuit during state I and switching sequence of one commutation state. In Figure 8a, when the switches S3 and S2 are turned ON, the back-EMF and battery voltage are connected in series to charge the armature inductor for building the armature current to generate the braking torque. Conversely, when the switches S3 and S2 are turned OFF, charges stored in the armature inductor are returned to the battery through the feedback diodes D4 and D1. As a result, energy regeneration and electric braking are achieved simultaneously. Since both the battery voltage and back-EMF are used to charge the armature inductor, the braking torque will be higher. In this method, when the motor comes to rest, the supply to the motor should be disconnected immediately. Otherwise, the motor will start to rotate in the reverse direction.

Three Switch Method

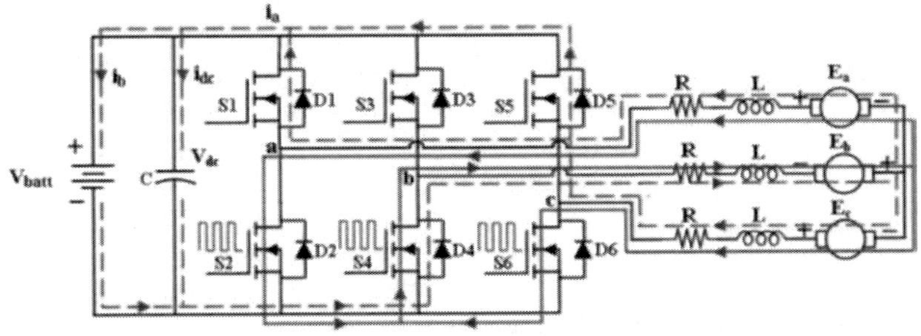

Figure 9. Equivalent Circuit of Three Switch method.

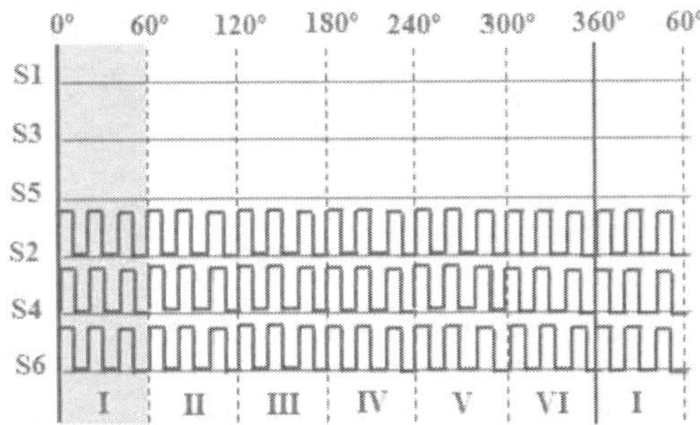

Figure 10. Switching Sequence of Three Switch Method.

In the three switch method, three switches are operated in each commutation state. The high side switches S1, S3 and S5 are always kept OFF, and the lower side switches S2, S4 and S6 are operated simultaneously in PWM switching mode [24, 25]. The operating principle of this method is similar to conventional passive dynamic braking, whereas the power resistors have been eliminated. However, the energy can be recovered. Figure 9 shows the equivalent circuit during state I and Figure 10 shows the switching sequence of three switch braking method. In state I, when the power switches S2, S4, S6 are turned ON, the back-EMF charges the armature inductor to generate the braking torque. In contrast, when the power switches are turned OFF, charges stored in the inductor are returned to the battery through the feedback diodes D1 and D5. As a result, both energy regeneration and electric braking are achieved simultaneously. Implementing the three switch method is easy since the rotor position information is not required for commutation.

Performance Evaluation of Single Stage Regenerative Braking Methods

The performance of single switch, two switch and three switch braking methods is carried out using both simulation and experiment. The performance of two parameters, namely stopping time and energy recovery, is studied. In performance evaluation, the characteristics of stopping time and energy recovery of each braking method are studied for various duty cycles from 0.1 to 1 with different speeds. The BLDC motor is subjected to run at a constant speed to apply brake signal for the performance analysis.

Simulation Results

The block diagram of the performance analysis study is depicted in Figure 11. The simulation model consists of a battery, three phase inverter, BLDC motor, and a control module. The control module is programmed with various braking methods such as single switch, two switch and three switch. The acceleration and brake commands are given to the control module that drives the motor for various driving conditions. When the acceleration command is applied, the brake command is made inactive and vice-versa. Simulation is carried out in Matlab/Simulink.

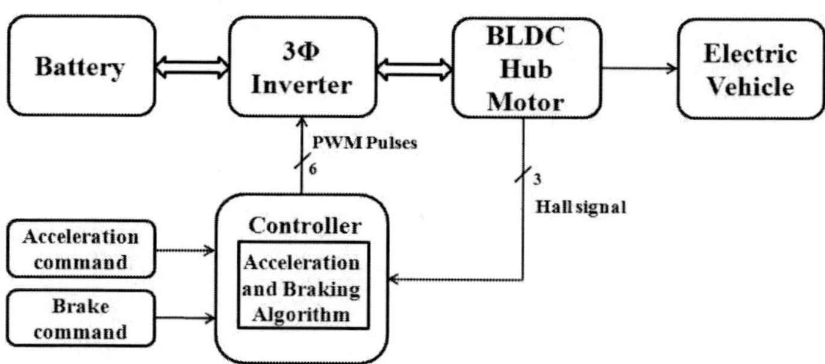

Figure 11. Block Diagram for Performance Analysis of Various Braking Methods.

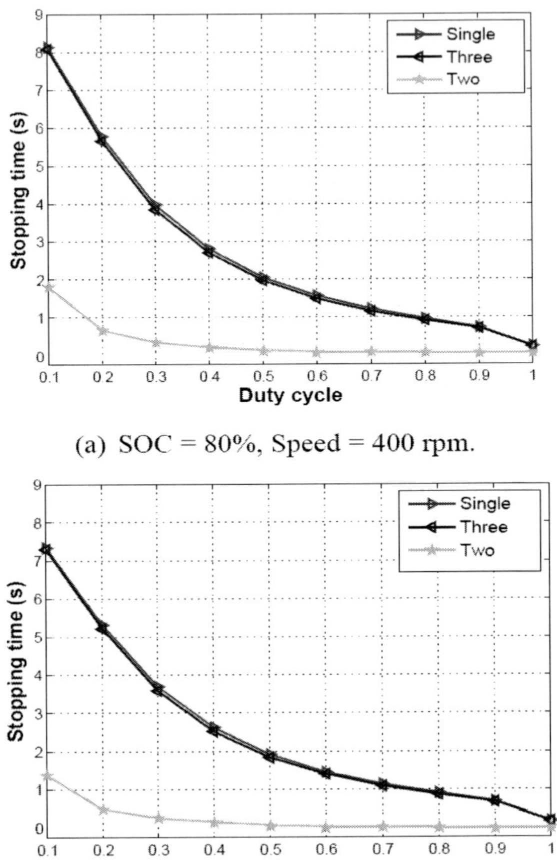

(a) SOC = 80%, Speed = 400 rpm.

(b) SOC = 80%, Speed = 200 rpm.

Figure 12. Simulation Results for Stopping Time Vs Duty cycle for SOC and Speed.

Stopping Time

Figure 12 shows the comparison of stopping time among the braking methods at various duty cycles with different speeds and 80% SOC levels. For a particular motor speed and SOC level as shown in Figure 12(a), the two-switch method exhibits a shorter stopping time, whereas the three-switch and single-switch methods result in longer stopping times. The stopping time of the three switch method is a little less than the single switch method. With the duty cycle increase, the stopping time of all the braking methods again decreases. In Figure 12(a) and 12(b), the stopping time is compared for the speed of 400 rpm and 200 rpm. Typically, the stopping time of a motor depends upon the speed at which it is running. It is observed that the stopping time of all the braking methods for 200 rpm consistently proves shorter than 400 rpm, and the pattern remains the same.

Energy Recovery

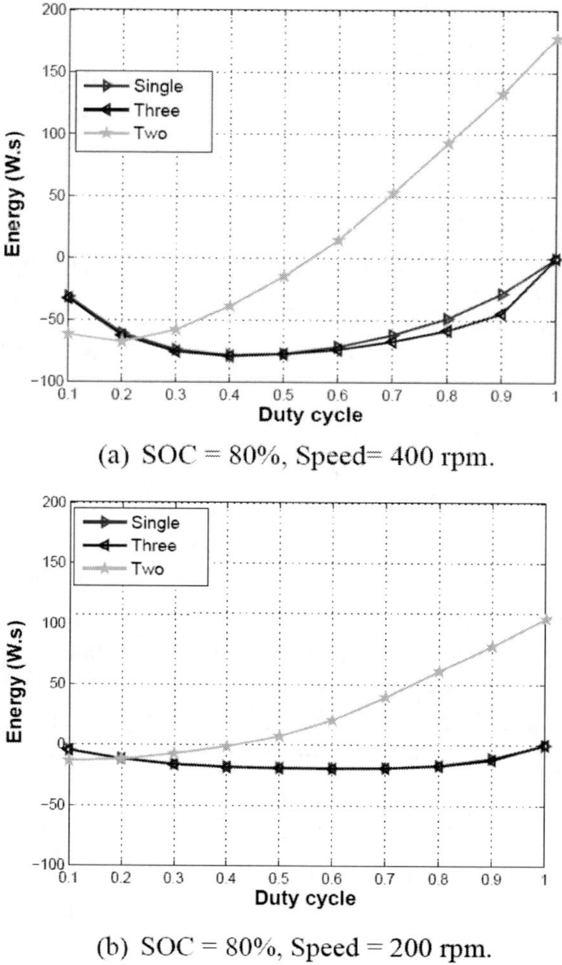

Figure 13. Simulation results for energy recovery Vs duty cycle for SOC and Speed.

The energy recovery during the braking period is compared for the same condition as done in stopping time analysis. In Figure 13, for the speed of 400 rpm and 80% SOC level, the energy recovered by three switch is higher among all the braking methods. The energy recovered in the single switch is slightly lesser than the three switch. As the duty cycle increases, the energy first increases to a peak value and then decreases and finally reaches zero when the duty cycle attains one. In two switch method, it is observed that the energy is negative during the duty cycle ranging from 0 to 0.5 and positive during 0.5 to 1. This is because that, during the duty cycle from 0 to 0.5, the amount of energy recovered is more than the amount of energy consumed from the battery over the braking period. Also, when the duty cycle ranges from 0.5 to 1, the amount of energy recovered is less than the amount of energy consumed from the battery, and it is positive. So in two switch, the effective regeneration occurs in the duty cycle range from 0 to 0.5. The effective energy recovered by two switch also increases to a peak value and then decreases until 0.5 duty cycle. At low duty cycles, the energy recovery by two switch is higher than single and three switch. While comparing with 200 rpm and same SOC, it is observed that the energy recovered and energy consumed for 200 rpm is less for all the methods and the pattern remains the same.

Experimental Results

The performance study of each braking method is conducted separately for a specific level of SOC and speed with various duty cycles similar to the simulation study. The motor is accelerated to run at a speed of 340 rpm with a 50% SOC level and subjected to braking after it attains steady speed. The stopping time of each braking method is examined from the speed waveform and energy recovery in the form of average current from the battery current waveform. The stopping time and average regenerative current over the braking period for various duty cycles are recorded and is shown in Table 1.

Table 1. Stopping time and energy recovery for various duty cycle

Duty cycle	Stop time (s)			Average current (A.s)		
	Single	Three	Two	Single	Three	Two
0.2	2.2	2.15	1.09	-0.112	-0.214	-0.189
0.4	1.49	1.39	0.5	-0.239	-0.247	-0.091
0.6	1.01	0.99	0.26	-0.286	-0.289	0.355
0.8	0.75	0.7	0.24	-0.234	-0.244	0.889

The table indicates that the stopping time of two switch is low, while the three and single switch exhibit comparatively longer stopping times across all duty cycles. Moreover, the stopping time of three switch is slightly lower than that of single switch method. As the duty cycle increases, the stopping time of each braking method decreases. The average regenerated current in three switch is slightly higher than the single switch for all the duty cycles. In two switch, regeneration occurs for the duty cycles less than 0.5 and no regeneration above 0.5. The experimental results match with the pattern of simulation results. Figure 14 shows the experimental results of single switch, two switch and three switch method taken for the duty cycles 0.4 and 0.8.

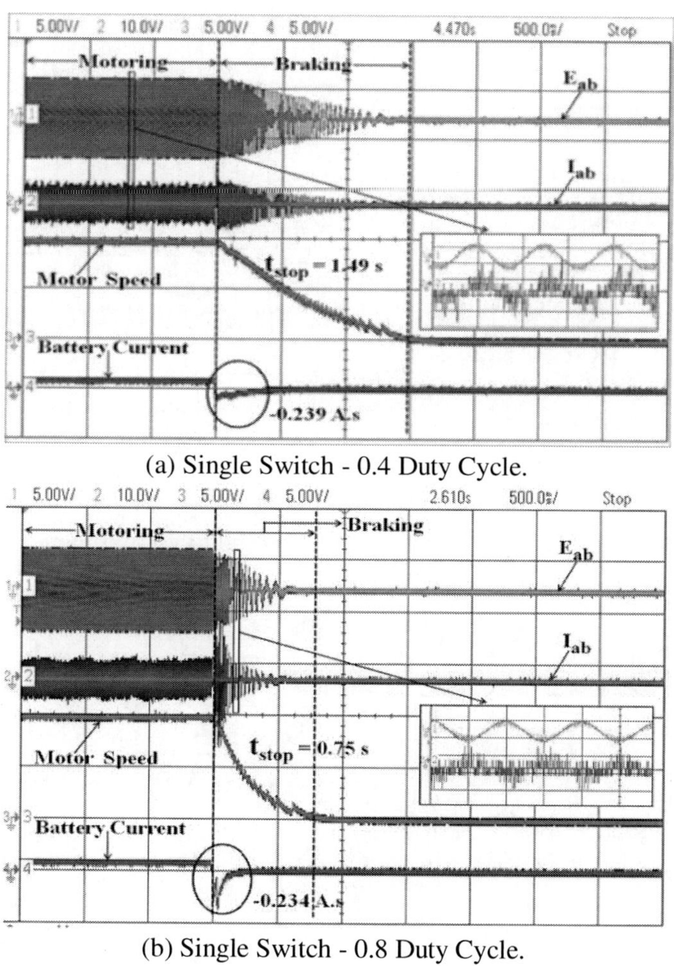

(a) Single Switch - 0.4 Duty Cycle.

(b) Single Switch - 0.8 Duty Cycle.

Figure 14. (Continued)

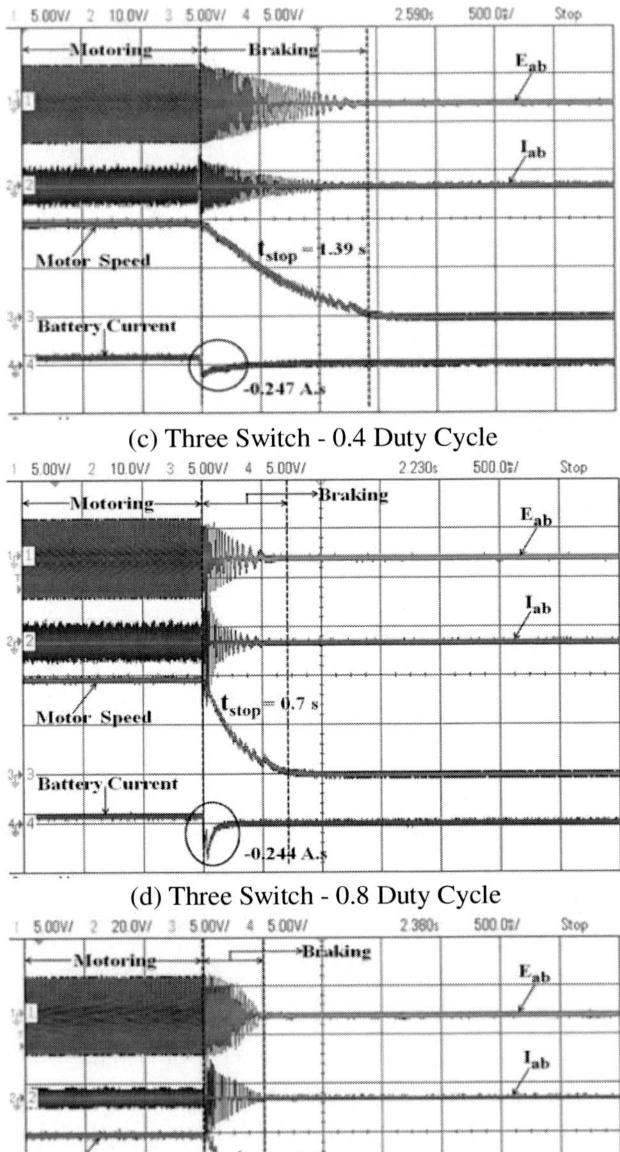

(c) Three Switch - 0.4 Duty Cycle

(d) Three Switch - 0.8 Duty Cycle

(e) Two Switch-0.4 Duty Cycle

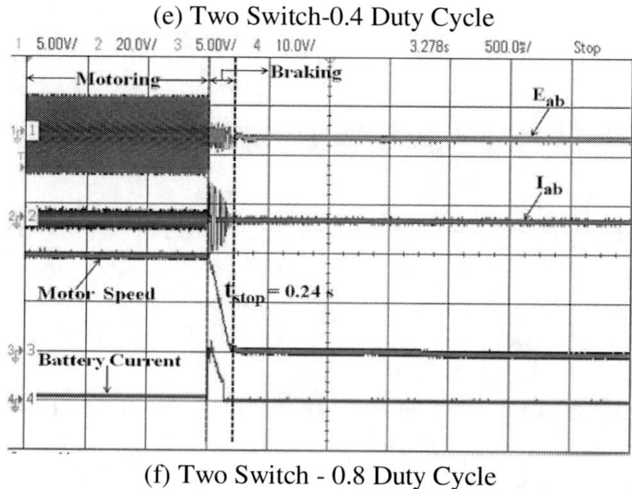

(f) Two Switch - 0.8 Duty Cycle

Figure 14. Experiment Results of Performance Evaluation for Single Switch, Two switch and Three Switch Under 50% SOC and 340 rpm with 0.4 and 0.8 duty cycle.

Future Trends

The use of more efficient BLDC motors, the development of new control strategies, the use of new materials, and the development of integrated systems are all emerging trends in regenerative braking technology. These trends have the potential to significantly impact the future of electric vehicles. By making regenerative braking systems more efficient and effective, these trends can help to extend the range of electric vehicles, reduce their charging time, and improve their overall performance.

For example, increasing the efficiency of regenerative braking systems by just 10% could increase the range of electric vehicles by up to 5%. This could make electric vehicles more viable for long-distance travel. Additionally, reducing the charging time of electric vehicles by just 10% could make them more convenient to use.

Overall, the emerging trends in regenerative braking technology have the potential to make electric vehicles more appealing to consumers and accelerate the adoption of electric vehicles.

In conclusion, regenerative braking is a valuable technology that can help to extend the range of electric vehicles and improve their energy efficiency. BLDC-driven electric vehicles typically have more efficient regenerative braking systems than other types of electric vehicles. There are a number of ways to enhance the effectiveness of regenerative braking in BLDC-powered electric vehicles.

By carefully considering the system design and addressing the implementation challenges, it is possible to create and implement regenerative braking systems that are safe, reliable, and effective.

The emerging trends in regenerative braking technology have the potential to significantly impact the future of electric vehicles. By making regenerative braking systems more efficient and effective, these trends can help to extend the range of electric vehicles, reduce their charging time, and improve their overall performance.

Overall, the emerging trends in regenerative braking technology have the potential to make electric vehicles more appealing to consumers and accelerate the adoption of electric vehicles.

Conclusion

In conclusion, regenerative braking is a valuable technology that can help to extend the range of electric vehicles and improve their energy efficiency. BLDC-driven electric vehicles typically have more efficient regenerative braking systems than other types of electric vehicles. There are a number of ways to enhance the effectiveness of regenerative braking in BLDC-powered electric vehicles.

By carefully considering the system design and addressing the implementation challenges, it is possible to create and implement regenerative braking systems that are safe, reliable, and effective.

The emerging trends in regenerative braking technology have the potential to significantly impact the future of electric vehicles. By making regenerative braking systems more efficient and effective, these trends can help to extend the range of electric vehicles, reduce their charging time, and improve their overall performance.

Overall, the emerging trends in regenerative braking technology have the potential to make electric vehicles more appealing to consumers and accelerate the adoption of electric vehicles.

References

[1] Zhang, Z., C. Wang, Y. Chen, H. Chen, S. Xu, and Y. Zhang, "Design and implementation of a regenerative braking system for a BLDC-driven electric vehicle," *IEEE Transactions on Power Electronics*, vol. 37, no. 11, pp. 12180-12194, 2022.

[2] Hu, J., B. Lu, J. Li, and S. Song, "A novel regenerative braking control strategy for BLDC motor-driven electric vehicles," *IEEE Transactions on Industrial Electronics*, vol. 68, no. 10, pp. 9557-9567, 2021.

[3] Zhang, Y., K. Wu, J. Hu, J. Li, and S. Song, "A fuzzy logic-based regenerative braking control strategy for BLDC motor-driven electric vehicles," *IEEE Transactions on Transportation Electrification*, vol. 6, no. 4, pp. 1411-1421, 2020.

[4] Zhang, D., Y. Lu, H. Wang, and Z. Wang, "A hybrid control strategy for regenerative braking of BLDC motor-driven electric vehicles," *IEEE Transactions on Vehicular Technology*, vol. 69, no. 10, pp. 10570-10582, 2020.

[5] Li, H., W. Song, and J. Wang, "A review of regenerative braking control strategies for electric vehicles," *Energies*, vol. 15, no. 16, p. 5984, 2022.

[6] Hannig, F., T. Smolinka, P. Bretschneider, S. Nicolai, S. Krüger, F. Meißner, and M. Voigt (2009) Stand und entwicklungspotenzial der speichertechniken für elektroenergie–ableitung von anforderungen an und auswirkungen auf die investitionsgüterindustrie. *Abschlussbericht, BMWi-Auftragsstudie*, 8, 28.

[7] Kubaisi, R. (2018). *Adaptive Regenerative Braking in Electric Vehicles (Doctoral dissertation, Dissertation, Karlsruhe, Karlsruher Institut für Technologie (KIT),* 2018).

[8] Liang, J., P. D. Walker, J. Ruan, H. Yang, J. Wu, and N. Zhang (2019) Gearshift and brake distribution control for regenerative braking in electric vehicles with dual clutch transmission. *Mechanism and Machine Theory*, 133, 1–22.

[9] Bian, Y., L. Zhu, H. Lan, A. Li, and X. Xu (2012) Regenerative braking strategy for motor hoist by ultracapacitor. *Chinese journal of mechanical engineering*, 25(2), 377–384.

[10] Li, L., X. Li, X. Wang, J. Song, K. He, C. Li, Analysis of downshifts improvement to energy e_ciency of an electric vehicle during regenerative braking, *Applied Energy* 176 (2016) 125-137.

[11] Zhe, L., Z. Ling, R. Yue, Y. Wei, L. Yinong, G. Feng, L. Yusheng, X. Zhoubin, A control strategy of regenerative braking system for intelligent vehicle, in: Intelligent and Connected Vehicles (ICV 2016), *IET International Conference on, IET*, 2016.

[12] Pan, C., L. Chen, L. Chen, H. Jiang, Z. Li, S. Wang, Research on motor rotational speed measurement in regenerative braking system of electric vehicle, *Mechanical Systems and Signal Processing* 66-67 (2016) 829-839.

[13] Kumar, M. S., S. T. Revankar, Development scheme and key technology of an electric vehicle: *An overview, Renewable and Sustainable Energy Reviews* 70 (2017) 1266-1285.

[14] Jeong, C. L., J. Hur, A novel proposal to improve reliability of spoketype bldc motor using ferrite permanent magnet, *IEEE Transactions on Industry Applications* 52 (5) (2016) 3814-3821.

[15] Kim, T. (2011) Regenerative braking control of a light fuel cell hybrid electric vehicle. *Electric Power Components and Systems*, 39(5), 446–460.

[16] Hegazy, O., J. Van Mierlo, and P. Lataire (2012) Analysis, modeling, and implementation of a multidevice interleaved dc/dc converter for fuel cell hybrid electric vehicles. *IEEE transactions on power electronics*, 27(11), 4445–4458.

[17] Khaligh, A. and Z. Li (2010) Battery, ultracapacitor, fuel cell, and hybrid energy storage systems for electric, hybrid electric, fuel cell, and plug-in hybrid electric vehicles: State of the art. *IEEE transactions on Vehicular Technology*, 59(6), 2806–2814.

[18] Armenta, J., C. Núñez, N. Visairo, and I. Lázaro (2015) An advanced energy management system for controlling the ultracapacitor discharge and improving the electric vehicle range. *Journal of Power Sources*, 284, 452–458.

[19] Yang, Y. P., J. J. Liu, T. J. Wang, K. C. Kuo, and P. E. Hsu (2007) An electric gearshift with ultracapacitors for the power train of an electric vehicle with a directly driven wheel motor. *IEEE Transactions on vehicular technology*, 56(5), 2421–2431.

[20] Yang, Y. P., J. J. Liu, and T. H. Hu (2011) An energy management system for a directly-driven electric scooter. *Energy Conversion and management*, 52(1), 621–629.

[21] Oleksowicz, S., K. Burnham, and A. Gajek (2012) On the legal, safety and control aspects of regenerative braking in hybrid/electric vehicles. Czasopismo Techniczne. *Mechanika*, 109(3-M), 139–155.

[22] Nian, X., F. Peng, and H. Zhang (2014) Regenerative braking system of electric vehicle driven by brushless dc motor. *IEEE Transactions on Industrial Electronics*, 61(10), 5798–5808.

[23] Yang, M. J., H. L. Jhou, B. Y. Ma, and K. K. Shyu (2009) A cost-effective method of electric brake with energy regeneration for electric vehicles. *IEEE Transactions on Industrial Electronics*, 56(6), 2203–2212.

[24] Chi, W. C., M. Y. Cheng, and C. H. Chen (2013) Position-sensorless method for electric braking commutation of brushless dc machines. *IET Electric Power Applications*, 7(9), 701–713.

[25] Dubey, M. K. and P. B. Bobba (2019) Variable switch regenerative braking technique for pm bldc motor driven electric two-wheeler. In *E3S Web of Conferences*, volume 87. EDP Sciences, 2019.

Chapter 4

Driving the Future: Civil Engineering's Integrated Role in Electric Vehicle Technology and Infrastructure

N. S. Padmavathy[1,*]
Mohit Hemath Kumar[2]
Rajcsh Kumar[2]
and R. Ruban[3]
[1]Department of Civil Engineering, National Institute of Technology, Tiruchirappalli, Tamil Nadu, India
[2]Department of Mechanical Engineering, Alliance College of Engineering and Design, Alliance University, Bengaluru, Karnataka, India
[3]Department of Mechanical Engineering, National Institute of Technology, Tiruchirappalli, Tamil Nadu, India

Abstract

The fast expansion of electric vehicles (EVs) is resulting in a noteworthy shift in our transportation environment. This abstract explores how important civil engineering is to the development of EV technology and the infrastructure that supports it. A reliable and effective network of charging stations is essential to the success of EV adoption, and charging station design, construction, and optimization are the domain expertise of civil engineers. They also play a vital role in developing sustainable materials for the construction of infrastructure and EV manufacturing. The impact of rising EV usage on our road infrastructure, including traffic management and road capacity, is the responsibility of civil

* Corresponding Author's Email: padma.struct@gmail.com.

In: Electric Vehicle Technology Structure, Instrumentation and Challenges
Editors: S. N. Sundaram, P. N. Sundaram, M. H. Kumar et al.
ISBN: 979-8-89113-695-3

engineers. In addition, this abstract highlight the necessity of multidisciplinary cooperation between environmental campaigners, engineers, and legislators to incorporate EVs into our current transportation networks smoothly. Civil engineers are leading the way in the development of electric vehicle technology and are essential to the sustainable transportation of the future.

Keywords: civil engineering, electric vehicles, sustainable materials, transportation systems

Introduction

We are living in a critical moment in the history of transportation. The combination of civil engineering and electric vehicle (E.V.) technologies has emerged as a ray of hope as urbanization grows and the environmental costs of traditional fossil fuel-based transportation become more evident. This introduction highlights the importance of sustainable transportation options in the modern environment. It provides a starting point for a deeper look at the complex interaction between E.V. technology and civil engineering (Hiep et al. 2023; Jain et al. 2023).

The Rise of Electric Vehicles

Electric vehicles are driving a significant revolution in the automobile sector. E.V.s, which run on electricity stored in large-capacity batteries, are becoming increasingly well-liked as an environmentally friendly and sustainable substitute for conventional internal combustion engine (ICE) automobiles. Numerous essential elements are driving this paradigm shift:

Environmental Concerns

E.V.s are becoming increasingly popular due to the pressing need to slow climate change and cut greenhouse gas emissions. In contrast to their gasoline and diesel counterparts, electric vehicles have zero exhaust emissions, which significantly reduces local air pollution and aids in the worldwide fight to combat climate change.

Developments in Battery Technology

E.V.s may now offer competitive ranges, quick charging times, and enhanced overall performance thanks to the creation of sophisticated lithium-ion batteries and other energy storage options. Not only have these developments increased the allure of E.V.s, but they have also created opportunities for new uses.

Government Regulations and Incentives

By providing incentives like tax breaks, rebates, and access to carpool lanes, numerous governments worldwide aggressively encourage the use of E.V.s. Concurrently, regulatory actions are driving automakers' production of more electric and hybrid vehicles by imposing tighter emissions criteria.

Consumer Demand

There is a growing interest in electric vehicles as consumers become more environmentally concerned and energy costs rise. Consumer interest in electric cars (E.V.s) has increased due to improved prices, convenience, and a more comprehensive range of models.

Technological Advancements

Incorporating state-of-the-art technology, like intelligent entertainment systems and autonomous driving capabilities, into E.V.s has significantly boosted their appeal. The transportation scene is changing due to the widespread use of electric vehicles, but the effects go well beyond the auto industry. The infrastructure required to facilitate the adoption of electric cars is being created in a way that redefines the function of civil engineering.

Transportation and Civil Engineering's Role

Roads, bridges, tunnels, airports, and public transportation systems are all part of the transportation infrastructure, and civil engineering is the foundation of it all. Historically, the construction of infrastructure to support gasoline and diesel-powered cars has been the primary goal of civil engineering in the transportation industry. But as the globe moves toward electric vehicles, civil engineering needs to change and advance to keep up. The integration of electric vehicle technology can be attributed to civil engineering in multiple essential areas:

Infrastructure for Charging

A vast network of charging stations must be established to encourage the use of electric vehicles. Civil engineers are responsible for planning, creating, and erecting charging stations at residences, offices, public parking spaces, and highways. This includes taking accessibility, urban planning, and electrical infrastructure into account.

Integration of Roadways and Highways

Roads and highways must be built to accommodate electric vehicles effectively. This involves creating specific lanes or corridors for electric cars and implementing innovative road technologies which can charge E.V.s while they travel.

Bridge and Tunnel Adaptations

Infrastructure may need to be modified to meet the unique requirements of electric vehicles. For example, bridges and tunnels that can safely carry the weight of E.V.s may need to be built. Civil engineers need to consider these factors while evaluating current structures and designing new ones.

Urban Planning for EV-Friendly Cities

The adoption of electric vehicles necessitates the development of public transit and infrastructure for charging. Civil engineers are essential to developing urban environments that support sustainable transportation and make it easier for E.V.s to access.

Sustainability in Transportation

Eco-friendly materials, energy-efficient design, and a more negligible environmental impact are all stressed in civil engineering methods increasingly centered on sustainability. The general objective of developing a more sustainable transportation system aligns with this (Mohammadi et al. 2023).

The Value of Environmentally Friendly Transportation

Sustainable transportation is more than just a catchphrase; it signifies a significant change in how we think about mobility and environmental stewardship. We must modify how we move people and products to combat climate change and reduce our carbon footprint.

Cutting Emissions of Greenhouse Gases

The desire to cut greenhouse gas emissions is one of the leading forces behind sustainable transportation. These emissions are caused mainly by the burning of fossil fuels in gasoline and diesel engines in the transportation sector. Electric cars offer a practical and quick fix for this issue since they run on power from greener sources.

Air Quality and Public Health

Concerns about air quality and public health are also addressed by sustainable mobility. Air pollution from conventional cars causes respiratory disorders and other health problems. We can drastically cut hazardous pollutants and enhance the air quality in our cities by switching to electric automobiles.

Resource Conservation

Preserving natural resources is a prerequisite for sustainable transportation. Because they need fewer natural resources to operate and are often more energy-efficient, electric vehicles help to reduce resource use and environmental deterioration.

Economic Benefits

Putting money into environmentally friendly transportation can pay off. An important driver of economic growth, innovation, and job creation is the electric vehicle (E.V.) sector. Less reliance on fossil fuels can also increase pricing stability and energy security.

The electrification of transportation also presents opportunities for global mobility solutions. E.V. technology can be used in various contexts, such as in underdeveloped nations where electrification can increase access to transportation and provide jobs. In conclusion, the fusion of electric vehicle technology with civil engineering is a reaction to the pressing issues of our day, not only technical or infrastructural issues. It signifies a change toward economic expansion, less emissions, better air quality, and sustainability (Ahmad and Satrovic, 2023).

Electric Vehicles (E.V.s) and Their Impact

A significant transformation is taking place in the transportation sector, which has the potential to transform how we travel and tackle some of the most critical environmental issues of our day. Electric vehicles (E.V.s), which provide a greener, more sustainable option to conventional gasoline and diesel-powered vehicles, have become a significant factor behind this shift. We'll get into the profound effects of electric cars in this introduction, focusing on the innovations in technology, the advantages for the environment, and the social ramifications of this revolution in transportation.

Over the past twenty years, the automotive industry has undergone a notable transformation. From being on the industry's periphery, electric vehicles are now at the forefront of transportation innovation. The following crucial elements have come together to fuel the growth of electric cars:

Environmental Imperative

An increasing understanding of the pressing need for ecological action lies at the core of the electric vehicle boom. The threat of climate change is growing, so reducing greenhouse gas emissions is essential. Due to their reliance on fossil fuels, conventional internal combustion engine (ICE) cars are well recognized for releasing hazardous pollutants such as carbon dioxide (CO2) into the atmosphere. Since electric vehicles are meant to have zero exhaust emissions, they answer this environmental problem honestly.

Developments in Battery Technology

The astounding advancements in battery technology are a crucial component of the viability of electric vehicles. More specifically, lithium-ion batteries' energy density, cost-effectiveness, and performance have all improved significantly. These developments have enabled electric cars to provide competitive driving ranges, rapid charging periods, and top-notch performance. The lithium-ion battery essentially acts as the vehicle's heart for an electric car because it can power the electric motor and offer a practical, emission-free driving experience.

Rules and Incentives from the Government

Promoting the usage of electric cars requires the support of all governments. Benefits, including tax rebates, refunds, and access to carpool lanes, are being offered to attract clients. Furthermore, manufacturers are compelled by law to produce more hybrid and electric cars to meet stricter emissions standards. The goal is to move toward cleaner modes of transportation as quickly as possible.

Customer Demand

The electric vehicle revolution is mainly driven by consumer demand. Electric vehicles have become increasingly popular as more people prioritize environmental sustainability, look to lower their energy expenses and embrace technological innovation. An increasing number of people are finding electric vehicles a desirable option due to their enhanced affordability, convenience, and expanded model choices.

Technological Developments

Electric cars use state-of-the-art technology and are environmentally beneficial substitutes for conventional automobiles. Their desirability has increased due to the addition of connectivity technologies, intelligent infotainment systems, and autonomous driving capabilities. Because of these technical developments, driving has been redefined, and electric vehicles have come to represent innovation and advancement. The ongoing spread of electric vehicles will significantly impact the transportation scene. This change changes society's perception of mobility and goes beyond the car industry. They are adapting and supporting electric mobility challenges in the automobile industry, civil engineering, and infrastructure development (Liu et al. 2023).

Environmental Benefits of E.V.

The clear environmental benefits of electric vehicles are the driving force behind their rise in popularity. These benefits go far beyond the lack of tailpipe emissions; they are an essential part of the more environmentally friendly and sustainable transportation ecosystem. Here, we highlight the benefits that electric cars have for the environment:

Zero Tailpipe Emissions

Particulate matter, nitrogen oxides, and carbon monoxide are just a few harmful pollutants that conventional gasoline and diesel cars release into the atmosphere. These pollutants not only worsen the air quality but also present serious health hazards, making respiratory issues and other conditions more likely. In sharp contrast, electric cars emit zero tailpipe emissions, which means that none of these dangerous compounds are released into the atmosphere. This is particularly important in cities, where air pollution frequently reaches hazardous levels, endangering the health of the general people.

Lower Greenhouse Gas Emissions

The main advantage of electric cars for the environment is their markedly lower greenhouse gas emissions. Because internal combustion engine cars burn fossil fuels, the transportation industry is one of the world's primary sources of carbon emissions. Carbon emissions from electric vehicles are significantly reduced, especially when the electricity is charged from clean and renewable sources. This decrease is essential to the worldwide effort to slow down climate change and limit the increase in global temperatures.

Energy Efficiency

Electric vehicles are intrinsically more energy-efficient than gasoline or diesel counterparts. In contrast to ICE vehicles, which lose much energy as waste heat, they convert a higher percentage of the energy from the grid into vehicle movement. This increased energy efficiency reduces overall energy consumption and, as a result, has a less negative effect on the environment.

Conservation of Resources

Electric vehicles use fewer resources and are more energy efficient. Crude oil must be extracted and refined, a resource-intensive and environmentally harmful process, to produce gasoline and diesel fuels. On the other hand, electric vehicles run on electricity, which can come from several sources, including renewable energy sources like hydropower, solar energy, and wind. This change lessens the need for oil and its adverse environmental effects.

Noise Reduction

Despite not affecting emissions, electric automobiles also reduce noise pollution. Conventional internal combustion engines are notorious for their loudness and vibrations, which may be troublesome in homes and cities. Electric cars provide a quieter and more understated form of mobility because of their innate silence.

In summary, electric cars are a revolutionary development in energy efficiency and environmental sustainability. Their rise is not just a fad but a calculated reaction to the three primary global issues of resource conservation, climate change, and air pollution. In addition to providing solutions for these issues, transportation electrification portends a more sustainable and clean future (Li and Wang, 2023).

Introduction : Electric Vehicle Technology

Electric vehicles (E.V.s) are a significant development in transportation because they offer cleaner, more sustainable travel. An overview of the many parts and characteristics of electric vehicles (E.V.s) is given in this section, along with information on the advancements in battery technology and the growth of the charging infrastructure. It offers entry into the complex world of electric car technology. The technology that powers our cars and transforms how we think about personal mobility is at the core of the electric vehicle revolution. Comprehending the fundamental elements and modes of operation of electric vehicles (E.V.s) is essential to appreciate this revolution's significance fully.

Motor Electric

An electric motor replaces an electric vehicle's conventional internal combustion engine. The car's wheels are powered by mechanical energy from the battery through electromagnetic induction. Driving with an electric motor is smooth and responsive because of its economy, quiet operation, and quick torque delivery.

Battery Pack

An electric vehicle's battery pack is an energy storage unit. Within lithium-ion batteries, electrical energy is stored as chemical energy. These cells are arranged into packs that differ in chemistry, size, and capacity. The battery pack is vital in electric vehicles (E.V.s) because its ability dictates its operating range.

Power Electronics

These systems regulate how electricity is transferred from the battery to the electric motor and other car parts. They control the power output, handle regenerative braking to collect energy during deceleration and convert the motor's battery-derived direct current (D.C.) into alternating current (A.C.).

Inverter

The inverter transforms battery-derived D.C. power into the A.C. electricity needed to run an electric motor. It is essential for regulating the motor's speed and torque, which enables exact control over the vehicle's performance.

Charging Connector

To connect to external power sources, including home chargers or charging stations, E.V.s come with a charging connector. The charging standards and speeds supported by different E.V. models range from Level 1 (standard household outlets) to Level 2 (unique charging stations) and Level 3 (D.C. rapid charging).

Onboard Charger

An essential part of an electric vehicle, the onboard charger transforms incoming grid power from A.C. to D.C. electricity that may be stored in the car's battery. It establishes the fastest rate at which an electric vehicle may be charged; bigger capacity chargers allow quicker charging.

Regenerative Braking System

Regenerative brakes are a feature of electric cars that help them regain energy as they brake or slow down. The electric motor runs in reverse to produce energy and recharge the battery when the driver depresses the accelerator or applies the brakes.

Gaining an appreciation of electric vehicles' efficiency and performance benefits requires understanding how these parts work. Electric cars use energy more directly and sustainably than conventional internal combustion engines (ICE), which lose energy through heat and mechanical inefficiencies (Martins et al. 2023).

Battery Technology and Advancements

The range, performance, and cost of an electric vehicle are significantly influenced by its battery, sometimes called the vehicle's "heart." To fully realize the promise and accelerate the rise of electric cars, one must be aware of the most recent advancements in battery technology.

Lithium-Ion Batteries

Most electric cars utilize lithium-ion batteries due to their high energy density, lightweight design, and long cycle life. Tremendous advancements in battery technology in recent years have reduced prices and increased energy storage capacity.

Solid-State Batteries

The development of solid-state batteries signifies a breakthrough in battery science. These batteries replace the solid electrolyte used in traditional lithium-ion batteries, providing higher energy density, faster charging, enhanced safety, and a longer lifespan. Even though they are still in the research and development stage, solid-state batteries offer immense promise for the development of electric automobiles.

Energy Density

As batteries' energy density rises, lighter, more compact battery packs with greater driving ranges are made feasible. Additionally, increases in energy density allow electric cars to be smaller yet have practical driving ranges.

Fast Charging

The method for charging an electric car is evolving as fast-changing technology develops. Owning an electric vehicle can be more convenient thanks to high-power charging stations, which can quickly deliver significant energy.

Battery Longevity

Lithium-ion batteries are becoming more durable as technology advances. As a result, owners of electric vehicles may anticipate more extended periods of good battery performance, ultimately lowering the overall cost of ownership.

Sustainability and Recycling

The electric car sector is currently looking into ways to recycle and reuse batteries at the end of their useful lives. This strategy reduces the harmful effects of battery production and disposal on the environment while promoting sustainability.

Comprehending these developments in battery technology contributes to deciphering the remarkable strides achieved in electric automobiles. It emphasizes the possibility of even more economical, ecologically friendly, and efficient E.V.s in the future (Sarmah et al. 2023).

Infrastructure for Charging

An essential component of the deployment of electric vehicles is the accessibility of infrastructure for charging. A strong and extensive charging network is necessary to alleviate "range anxiety" and establish electric vehicles as a viable option for daily commuting.

Level 1 Input

Standard household outlets are used for Level 1 charging, sometimes called trickle charging. Although it is the slowest charging technique, it is practical for charging overnight at home. Typically, this degree of charging increases the range by two to five miles per hour.

Level 2 Charging

Workplaces, public charging stations, and residential garages are frequently equipped with Level 2 charging stations. They are appropriate for most everyday charging demands because they have faster charging speeds and can add between 10 and 30 miles of range each hour of charging.

DC Fast Charging (Level 3)

To enable quick charging, D.C. fast charging facilities are usually found in cities and beside roads. These stations may quickly increase an electric vehicle's range significantly, increasing the viability of long-distance driving.

Home Charging Options

Many people who own electric vehicles install Level 2 charging stations in their garages as their home charging options. This lessens the need for public charging infrastructure while offering everyday charging convenience.

Public Charging Network

Public charging networks must expand for long-distance and convenient urban charging. With the growing prevalence of public charging stations in shopping malls, metropolitan centers, and highway regions, owning an electric vehicle is becoming more feasible for a wider variety of consumers.

Smart Charging

Intelligent charging solutions are starting to emerge to maximize energy efficiency and lessen the burden on the grid during moments of peak demand. With the help of these systems, consumers may take advantage of lower electricity costs, plan charging for off-peak times, and support grid stability.

Wireless Charging

Currently under development, wireless charging technology, sometimes called inductive charging, provides easy charging without needing physical wires. This technology is expected to play a significant role in the future of electric car charging.

As E.V.s become more and more common, the infrastructure for charging them must be developed to ensure that owners of these cars have access to convenient and reliable charging options. In addition to promoting a greater use of electric vehicles, the growth of this network is altering our understanding of vehicle refueling. Understanding the ins and outs of electric car technology, as well as advancements in battery science and the creation of charging infrastructure, is essential to appreciating the incredible journey that these vehicles have gone through. Together, these elements define the character and potential of electric automobiles and are crucial to comprehend how they will transform the transportation sector. As we continue to investigate this fascinating realm, we will discover more about the nuanced impacts of electric vehicle technology on society, infrastructure, and our views and experiences with transportation (Patil et al. 2023).

Sustainable Transportation and Civil Engineering

Integrating sustainable practices with civil engineering has become essential in the ever-evolving transportation sector. This introduction aims to set the stage for a discussion of how civil engineering will affect transportation in the future by highlighting the significance of sustainable transportation practices and how they are transforming our environment. Civil engineering has long been the cornerstone of the transportation infrastructure, handling the planning, designing, constructing, and maintaining of highways, bridges, tunnels, airports, and public transit systems. Sustainable transportation is entering a new era, and civil engineering is essential to this paradigm shift.

Infrastructure for Charging

Establishing a massive network of charging stations is necessary to promote the usage of electric automobiles. Civil engineers are responsible for designing, building, and installing charging stations at homes, workplaces, public parking lots, and along roads. This entails considering electrical infrastructure, accessibility, and urban planning.

Integration of Roadways and Highways

Electric cars must be able to travel on roads and highways designed to suit their needs. To achieve this, dedicated lanes or passageways for E.V.s must be constructed. Additionally, intelligent road technologies that enable E.V. charging while in motion must be implemented.

Bridge and Tunnel Adaptations

Infrastructure may need to be changed to accommodate the unique needs of electric cars. For example, bridges and tunnels that can safely carry the weight of E.V.s may need to be built. Civil engineers need to consider these factors while evaluating current structures and designing new ones.

Urban Planning for EV-Friendly Cities

The adoption of electric vehicles necessitates the development of public transit and infrastructure for charging. Civil engineers are essential When planning and constructing urban landscapes that support environmentally friendly transportation and convenient access for electric vehicles.

Sustainability in Transportation

Eco-friendly materials, energy-efficient design, and a more negligible environmental impact are all stressed in civil engineering methods increasingly centered on sustainability. The general objective of developing a more sustainable transportation system aligns with this (Bao et al. 2023).

The Importance of Sustainable Transportation Practices

"Sustainable transportation" refers to a paradigm shift in how we view mobility and environmental responsibility, not just a trendy phrase. We must modify how we move people and products to combat climate change and reduce our carbon footprint.

Reducing Greenhouse Gas Emissions

The need to cut greenhouse gas emissions is one of the main factors promoting sustainable transportation. These emissions are caused mainly by the burning of fossil fuels in gasoline and diesel engines in the transportation sector. Electric cars offer a practical and quick fix for this issue since they run on power from greener sources.

Air Quality and Public Health

Concerns about public health and air quality are also addressed by sustainable transportation. Air pollution from conventional cars causes respiratory disorders and other health problems. We can drastically cut hazardous pollutants and enhance the air quality in our cities by switching to electric automobiles.

Resource Conservation

Preserving natural resources is a prerequisite for sustainable transportation. Because they need fewer natural resources to operate and are often more energy-efficient, electric vehicles help to reduce resource use and environmental deterioration.

Economic Benefits

Putting money into environmentally friendly transportation can pay off. An important driver of economic growth, innovation, and job creation is the electric vehicle (E.V.) sector. Less reliance on fossil fuels can also increase pricing stability and energy security.

Global Mobility Solutions

There are chances for global mobility solutions as a result of electrifying transportation. E.V. technology can be used in various contexts, such as in underdeveloped nations where electrification can increase access to transportation and provide jobs. In summary, the convergence of sustainable mobility methods and civil engineering is a reaction to the pressing issues of our day, not merely a technical or infrastructure issue. It signifies a change toward economic expansion, less emissions, better air quality, and sustainability (Senapati et al. 2023).

EV-Friendly Roadways and Highways

The need for EV-friendly roads and highways is growing as electric vehicles (E.V.s) become more popular in the automotive industry. This section introduces the idea of intelligent roads—road networks specifically intended to accommodate electric vehicle needs—and how E.V. charging stations can be integrated into existing networks. Making E.V. charging stations a part of the road network is a critical step in transforming electric vehicles into a valuable and convenient form of transportation. This project is essential because it is more than just physical development; it is a fundamental change in how we think about and use our roads.

Strategic Location of Charging Infrastructure

The efficiency of E.V. charging stations depends on their location. These stations should be strategically positioned along main thoroughfares, highways, and metropolitan areas. They should provide various charging choices, such as fast chargers for expedient top-ups and Level 2 chargers for lengthier stays.

Highway Corridors

Long-distance and highway traffic routes are given priority in EV-friendly road networks. With a network of fast-charging stations installed along these corridors, owners of electric vehicles may comfortably take their cars on lengthy trips without worrying about running out of gas.

Urban Integration

E.V. charging infrastructure must be integrated into urban areas. This includes setting up charging stations at retail malls, commercial districts, and public parking lots. It also covers parking on the street. E.V. owners should be able to easily reach these charging stations while going about their everyday business.

Compatibility and Standardization

Industry-standard connectors and payment methods must be adopted to guarantee a smooth experience for E.V. drivers. Widespread acceptance and use are facilitated by consistency in the design and accessibility of charging infrastructure.

Renewable Energy Integration

By adding renewable energy sources like solar or wind turbines, E.V. charging stations can further promote sustainability. This method guarantees a cleaner energy supply for E.V.s while also lessening the environmental impact of charging.

Public-Private Partnerships

Funding, development, and upkeep of E.V. charging infrastructure sometimes needs cooperation between public and private groups. Establishing public-private partnerships can assist in addressing the financial obstacles linked to constructing an extensive network.

Smart Roads for E.V.s

The next stage of infrastructure development for transportation is smart highways for electric vehicles. These roads are made especially for electric cars, providing cutting-edge solutions that boost driving enjoyment, increase energy economy, and lessen environmental effects.

Dynamic Wireless Charging

EVs can be charged while driving thanks to integrating dynamic wireless charging technology into intelligent highways. With this technology, which embeds charging coils in the road surface, E.V.s that are compatible with it can charge continually while driving.

Vehicle-to-Grid (V2G) Integration

Smart roadways can help with V2G integration, allowing E.V.s to contribute extra energy back into the grid and draw electricity from it. Because of this two-way flow of electricity, E.V.s can become essential tools in energy management and help stabilize the grid.

Roadway Sensors and Communication

EV-interactive sensors and communication systems can be incorporated into intelligent roadways. These sensors can help with autonomous driving and monitor traffic and road conditions. This kind of communication improves traffic control and safety.

Real-Time Traffic Management

Smart roads can provide real-time traffic management solutions by gathering and evaluating data from E.V.s and the road infrastructure. This includes dynamic speed restrictions, adaptive traffic signals, and routing suggestions to lessen congestion and enhance overall traffic flow.

Enhanced Safety Features

Sensors and communication technologies that improve safety can be installed on smart roads. They can, for instance, identify potential dangers on the street, alert E.V.s to unfavorable weather, and even facilitate communication between moving cars to prevent collisions.

Energy-Efficient Lighting

Energy-efficient LED lighting that modifies its intensity in response to traffic conditions can be installed on smart roads. This lowers energy use while also increasing driving safety and visibility.

Environmental Sustainability

One of the main components of intelligent roads is using environmentally friendly materials and construction techniques. This entails putting green infrastructure beside roads, utilizing recycled materials, and improving road geometry for energy efficiency.

An essential step toward developing effective, secure, and environmentally friendly transportation in the future is the integration of intelligent highways for electric vehicles. It highlights the possibility of a comprehensive and networked transportation infrastructure that can support electric cars and improve their performance while lessening their environmental effect. In conclusion, the idea of smart roads and EV-friendly roads is not just a reaction to the increasing number of electric cars on the road; instead, it is a calculated step toward a more efficient and sustainable form of transportation in the future. The development of smart roads for E.V.s and the integration of E.V. charging facilities into road networks are progressively addressing our changing mobility needs.

Bridge and Tunnel Infrastructure for E.V.s

It is more important than ever to upgrade and modify our bridge and tunnel infrastructure to accommodate electric cars (E.V.s), which are becoming increasingly common on our roads. This introduction investigates how bridges and tunnels may be reimagined to make room for electric vehicles (E.V.s) and the structural factors needed to build infrastructure that supports E.V.s. For a long time, bridges and tunnels have been vital parts of our transportation network, allowing people and goods to move between places. But as the automotive industry transitions to electric mobility, these essential structures need to be rethought and modified to meet the unique needs of electric cars.

Weight-Bearing Capacity

This is the first and most crucial factor to consider when modifying bridges and tunnels for electric vehicles. The weight of the battery packs in electric cars is frequently more than that of conventional ICE vehicles. It is crucial to reinforce these structures to support this extra weight.

Height Clearance

Because of the batteries' location, electric cars—especially bigger SUVs and trucks—may have different heights than their internal combustion engine counterparts. Ample height clearance on bridges and tunnels is necessary to keep traffic flowing smoothly and prevent accidents.

Lightweight Materials

Using eco-friendly and lightweight building materials can improve the sustainability and efficiency of E.V. bridges and tunnels. These structures can support less weight overall and have less environmental impact when made of lightweight materials.

Integrated Charging Infrastructure

As E.V. usage rises, it's critical to have charging infrastructure within or close to bridges and tunnels. Level 2 or Level 3 charging stations should be a part of this integration to provide travelers with a practical way to charge their cars while on the go.

Energy-Efficient Lighting

Energy-efficient, EV-compatible lighting options are now available for bridges and tunnels. Intelligent lighting systems can save energy and improve visibility by adjusting their intensity in response to traffic conditions.

Durability and Weather Resistance

Extreme heat, cold, and dampness are just a few weather-related issues that bridges and tunnels must face. Maintaining the integrity and safety of these structures depends on ensuring their durability.

Structural Considerations for EV- Supportive Infrastructure

Careful attention to many structural factors is necessary when adapting bridge and tunnel infrastructure for electric cars. These factors cover the construction, materials, upkeep, and general robustness of these vital links in our transportation system.

Design of Bridges and Tunnels

New designs for bridges and tunnels need to consider the heavier weight of electric vehicles. To maintain the safety and structural integrity of these constructions, structural features that can support this weight must be incorporated by engineers.

Material Selection

The durability and sustainability of tunnels and bridges for electric cars greatly depend on the selection of building materials. Environmentally friendly, long-lasting, and lightweight modern building materials should be prioritized.

Retrofitting and Reinforcement

Existing tunnels and bridges may need to be reinforced to meet the structural requirements of electric vehicles. This could entail modifying height clearances, supporting foundations, and strengthening load-bearing components.

Regular Inspections and Maintenance

To detect deterioration, wear and tear, or other structural problems that would jeopardize these structures' safety, regular inspections and maintenance are crucial. Regular inspections might stop possible issues before they get out of hand.

Weatherproofing and Drainage

To shield bridge and tunnel infrastructure from the damaging effects of moisture, water, and temperature changes, adequate weatherproofing and drainage systems are necessary. These systems need to be built to resist changing weather patterns.

Advanced Monitoring Systems

By putting in place these systems, one may monitor the state of tunnels and bridges in real-time. These technologies allow for prompt intervention by identifying early stress, injury, or structural degeneration indicators. To improve our transportation network's safety, sustainability, and effectiveness, bridge and tunnel infrastructure must be modified to accommodate electric vehicles. This is more than just adding a new means of transportation. Electric vehicles are the way of the future for mobility because of their lower emissions and environmental advantages. Modern cars must easily integrate into our existing infrastructure, maintaining and improving the functionality and safety of these vital transportation components. To that end, we must reinvent our bridges and tunnels to make room for them.

Urban Planning and Smart Cities

The rise of electric vehicles (E.V.s) offers a singular chance to design smarter, more sustainable, and future-ready urban landscapes in the ever-evolving urban development and planning field. This introduction lays the groundwork for our investigation of the sustainable urban transportation models propelling change and the construction of smart cities with smooth E.V. integration. Intelligent cities entail utilizing technology, data, and inventive approaches to enhance the standard of living for inhabitants while mitigating their ecological footprint. Integrating electric vehicles into the urban fabric is vital to this concept, changing how people move, live, and interact in cities.

E.V. Charging Infrastructure

Creating an extensive E.V. charging infrastructure is fundamental to developing intelligent cities with E.V. integration. Ideally, charging stations should be positioned in urban areas so electric vehicle owners can easily access charging. In addition to encouraging the use of electric vehicles, this lessens range anxiety, allowing locals to transition to more environmentally friendly modes of transportation.

Multimodal Mobility

Smart cities give citizens various ways to move about by prioritizing multimodal mobility. Examples are programs for renting out electric vehicles, electric scooters, bikes, and reliable public transportation. Combining these choices promotes less car ownership, less traffic, and a more negligible environmental impact.

Connected Infrastructure

As E.V. usage increases, there is a need for interconnected infrastructure that enables smooth communication between E.V.s, charging stations, and traffic control systems. Grid-balancing techniques, real-time traffic management, and dynamic routing are made possible by this.

Urban Planning for E.V. Adoption

E.V. adoption needs to be considered in urban planning. This includes zoning regulations requiring charging infrastructure in newly constructed buildings, public spaces with E.V. chargers, and urban planning promoting cycling, walking, and electric vehicles.

Data-Driven Mobility

Smart cities use data to make transportation more efficient. To give locals real-time information they need to make educated transportation decisions, this may entail the deployment of sensors to monitor traffic, weather, and road conditions.

Sustainable Urban Transportation Models

A vital component of the idea of smart cities is the shift to environmentally friendly urban transportation technologies. These models prefer eco-friendly and economical mobility options consistent with sustainable development and intelligent expansion.

Public Transit Expansion

An essential component of sustainable urban transportation models is developing and improving public transit networks. Investing in high-capacity transportation choices like commuter trains, buses, trams, and subways is part of this.

Active Transportation

Walking and cycling are two forms of active transportation that sustainable urban transport promotes. Creating bike lanes, shared walkways, and pedestrian-friendly infrastructure encourages eco-friendly and healthful transportation choices.

EV-Sharing and Ride-Hailing

Implementing electric ride-hailing services and EV-sharing programs offers inhabitants practical and environmentally friendly transportation options. These services can improve traffic congestion, cut emissions, and decrease the number of people who own cars.

Transit-Oriented Development

Compact, mixed-use urban zones that minimize the need for private automobile traffic are encouraged by designing cities around transit hubs. Residents may walk, bike, or take public transportation to work, shopping, and entertainment.

Urban Mobility Hubs

These hubs serve as central hubs for integrating different forms of transportation, such as bicycles, public transit, and E.V. sharing programs. These hubs promote multimodal, sustainable travel by acting as easy transfer locations for commuters.

Electrification of Public Transportation

Using electric buses and other vehicles in public transportation fleets has a lot of environmental advantages. Electric buses have no tailpipe emissions, which helps clean up the air in cities.

Car-Free Zones

Designating areas in city centers as restricted or no-parking zones encourages bicycle and pedestrian traffic, eases traffic, and improves the urban environment.

Data-Driven Transportation Management

To improve traffic flow, lessen congestion, and boost efficiency, data-driven transportation management systems leverage real-time information. This method is essential for tackling the particular difficulties associated with urban travel.

Models of sustainable urban transportation acknowledge the significance of clean energy, many modes of transportation, and effective urban design.

Cities may lower their carbon footprint, clean up the air, improve the standard of living for citizens, and design the future of urban living by adopting these ideas. In conclusion, implementing sustainable urban transportation models and creating smart cities that include E.V.s constitute a comprehensive strategy for urban development. It involves creating intelligent, effective, and ecologically conscious cities.

Public Transport and Electric Mobility

A new era of cleaner, more efficient, and sustainable transit options is being ushered in by integrating electric mobility into the changing public transportation scene. The context for our inquiry into the electrification of public transportation and the integration of electric vehicles (E.V.s) with high-speed rail networks is provided by this introduction. Public transportation, which has always been seen as an essential component of urban mobility, is undergoing a significant change due to electrifying its fleet. This paradigm shift promises to reduce emissions while also providing an opportunity to reimagine the entire public transit experience.

Electric Buses

Electric buses are often the first vehicles to be electrified in public transportation. Electric buses have several advantages, such as less noise pollution, lower running costs, and no tailpipe emissions. They are instrumental in urban environments where air quality and noise pollution are significant problems.

Infrastructure for Charging

The creation of infrastructure for charging electric buses is essential as public transportation companies make the switch. To maintain seamless transit services, this infrastructure must meet the high energy requirements of electric buses and provide quick and dependable charging options.

Fleet Electrification

Buses are not just switching to electric mobility. Many public transportation organizations are investigating electrifying every vehicle in their fleet, including commuter rail systems, trolleys, and trams. By electrifying various transportation options, greenhouse gas emissions are decreased, and urban air quality is enhanced.

Battery Technology

The viability of electrified public transportation depends on the adoption of cutting-edge battery technology. Electric buses and trains must have large capacity, long-lasting batteries to run smoothly all day.

Sustainability Initiatives

Public transportation electrification frequently complies with larger environmental objectives. Some examples are using renewable energy sources to power transport systems, installing energy-efficient charging infrastructure, and encouraging sustainable behaviors across the transit sector.

Accessibility and Inclusivity

Electrification offers a chance to improve the accessibility and inclusivity of public transportation. Contemporary electric buses and trams frequently have low floors, ramps, and accommodations for patrons with limited mobility.

Urban Integration

Electric public transit must be incorporated into urban planning. Public transportation systems need to be made to be user-friendly, accessible, and have convenient routes to meet the needs of the community (Dawson et al. 2023; Silva et al. 2023).

High-Speed Rail and E.V. Integration

High-speed rail (HSR) is a critical component of contemporary transportation because it provides long-distance travel that is quick, economical, and ecologically benign. E.V. and HSR integration is one component of a multimodal strategy to increase the efficiency and sustainability of transportation.

Complementary Transportation Modes

E.V.s and high-speed rail are complementing forms of transportation. While E.V.s offer flexibility for local travel inside urban areas and regions not directly covered by HSR, HSR offers quick long-distance transport between major cities.

Last-Mile Connectivity

HSR stations are frequently found in the periphery of towns or cities. Passengers may reach their final destinations with the help of electric vehicles, which offer an appropriate solution for last-mile connectivity.

Intermodal Stations

HSR stations can act as centers for intermodal transportation, allowing passengers to move between trains and electric cars quickly. To improve the travel experience, these hubs may include EV-sharing services and infrastructure for charging.

Reduced Emissions

High-speed rail is renowned for its low carbon emissions and energy efficiency. Passengers can travel to and from HSR stations entirely emission-free thanks to the integration of E.V.s into the transportation network.

Charging Infrastructure at Stations

HSR stations may have E.V. charging infrastructure, enabling passengers to use high-speed rail services while their electric cars are set. This infrastructure allows drivers to drive fully charged electric vehicles farther on longer trips.

Reduced Congestion

Combining HSR with E.V.s can lessen airport and highway traffic. When traveling large distances, passengers can use high-speed trains, which avoids crammed roads and airports.

Environmental Benefits

There are a lot of environmental advantages to the joint use of E.V.s and HSR. It helps create a more sustainable transportation system by lowering the overall carbon footprint of long-distance travel.

Economic Opportunities

Combining HSR with E.V.s opens up tourism, employment growth, and regional development business prospects. Local economies benefit from the development of E.V. charging infrastructure and HSR construction.

Combining electric mobility with high-speed rail symbolizes a comprehensive, environmentally friendly, and effective transportation strategy. It improves travel quality, lowering emissions, boosting connections, and spurring economic growth. Conclusively, the combination of electric vehicles and high-speed rail, along with the electrification of public transit, signify noteworthy advancements towards a transportation network that is more efficient and sustainable. The way we travel, connect, and communicate within cities and over great distances is changing due to these advancements (Karakus et al. 2023).

Case Studies and Best Practices

This introduction centers on the noteworthy accomplishments of numerous projects and activities related to integrating electric vehicles (E.V.s) and the essential lessons that have been discovered. Through an analysis of these case studies and best practices, we may learn what has and has not succeeded and how to go forward in the rapidly changing field of EV-ready civil engineering. Examining effective programs and projects that include electric vehicles in our transportation system shows that advancement, creativity, and constructive change are possible. These case studies show the transforming impact of electrified mobility and serve as beacons of hope.

Urban E.V. Charging Networks

Several cities across the globe have started large-scale initiatives to establish comprehensive urban E.V. charging networks. Examples from cities such as Oslo, Norway, and Amsterdam, Netherlands, demonstrate how the thoughtful placement of infrastructure for charging EVs and adoption incentives can result in a notable rise in the usage of these vehicles.

High-Speed Rail Systems

The potential for a smooth integration with electric transportation is demonstrated by developing and growing high-speed rail systems in nations like France and Japan. In addition to cutting travel times, these projects have improved passenger convenience by using electric cars as first- and last-mile connectivity options.

Public Transit Electrification

Several successful electrification projects, with Shenzhen, China, setting the standard. These cities have achieved remarkable reductions in air pollution by switching to electric power for their entire fleet of buses, and they are now role models for other towns wishing to implement clean transportation solutions.

Smart City Initiatives

The promotion of electric mobility can be facilitated by sophisticated data-driven infrastructure, as demonstrated by case studies of smart cities such as Singapore and Barcelona, Spain. These cities optimize traffic flow, lessen congestion, and promote the use of electric vehicles and shared mobility options by fusing real-time data with intelligent transportation systems.

Electric Vehicle Fleets in Corporate Settings

Companies like UPS and Amazon have effectively implemented last-mile deliveries using electric vehicle fleets. These programs not only save emissions but also demonstrate the viability of using electric cars for business purposes, providing models for other companies wishing to switch to greener modes of transportation.

Public-Private Collaborations

Several prosperous initiatives have resulted from partnerships between public and private organizations. For example, E.V. charging corridors have been developed along critical roads in California due to public-private collaborations; this development provides a model for future intergovernmental cooperation in E.V. infrastructure.

Lessons Learned in EV-Ready Civil Engineering

There are obstacles and successes in the path to EV-ready civil engineering. Lessons from the past provide priceless direction for making future moves effectively and efficiently.

Scalable Infrastructure

Establishing a scalable infrastructure for E.V. charging is a crucial learning. The infrastructure must be adaptable enough to handle the growing demand for electric vehicles without necessitating expensive and disruptive overhauls as the market speeds up.

Standardization

Standardizing charging ports and payment methods is essential to guarantee a smooth and user-friendly experience for E.V. owners. The necessity for industry harmonization is highlighted by lessons from areas where standardization has been successful.

Integration with Urban Design

Promising case studies highlight how crucial incorporating E.V. infrastructure with urban design is. Zoning rules, building codes, and transportation planning must be created to encourage EV-ready cities with electric mobility in mind.

Adaptive Regulation

One lesson from areas where regulatory obstacles hindered the introduction of E.V.s is the need to modify regulatory frameworks to allow for electric mobility. Progressive laws can encourage infrastructure development, stimulate innovation, and encourage the use of electric vehicles.

Incentive and Education

Financial incentives, tax breaks, and public awareness campaigns are essential to promoting electric vehicle use (E.V.s). These insights come from areas that have successfully encouraged the adoption of E.V.s.

Battery Technology

One of the most critical lessons in electric transportation is the ongoing advancement of battery technology. Electric vehicles are becoming more feasible and accessible as batteries become more affordable, long-lasting, and energy dense.

Public-Private Partnerships

The effectiveness of initiatives fueled by these alliances serves as a reminder of the value of cross-sector cooperation. The lessons learned here show that infrastructure development can be accelerated, and service offerings can be enhanced by merging both sectors' resources, knowledge, and inventiveness.

Sustainability and Resilience

The significance of building infrastructure that can resist environmental challenges and promote a greener future is highlighted by lessons learned from projects that have integrated sustainability and resilience concepts.

Essentially, the analysis of accomplished projects and the knowledge gained from EV-ready civil engineering highlight the revolutionary potential of electric transportation and the demand for flexible, progressive approaches. By taking cues from other people's successes and experiences, we may effectively handle the opportunities and problems brought about by the electric vehicle revolution.

Challenges and Future Prospects

Although there are many obstacles to overcome, the story of invention and revolution at the nexus of electric vehicles (E.V.s) and civil engineering is fascinating. This introduction addresses the barriers that must be overcome for E.V.s to seamlessly integrate into our infrastructure and the bright future possibilities ahead. Introducing electric vehicles into our transportation system has ushered in a new era of mobility, spurred innovation, and lowered pollution. However, a few significant obstacles in this voyage need to be overcome.

Charging Infrastructure

Establishing a thorough and convenient charging infrastructure is one of the main issues facing E.V. users. For E.V. adoption to be widely adopted, charging stations must be installed in residences, places of business, public spaces, and along highways. It is still urgently necessary to overcome the finance, standardization, and scalability challenges.

Range Anxiety

The worry that an E.V. won't have enough battery life to reach a charging point remains a barrier to E.V. adoption. Technological developments in batteries and fast-charging networks are necessary to lessen this.

Battery Technology

Despite its quick advancement, battery technology still has issues with cost, recycling, and energy density. The industry's top goal is to develop more effective, durable, and ecologically friendly batteries.

Charging Time

One enduring issue is how long it takes to charge an electric car instead of filling it up with gas. One of the most critical areas for research and development is shortening setting periods and creating fast-charging systems.

Infrastructure Compatibility

It might be challenging to make E.V.s and the current infrastructure compliant with electrical grids, voltage levels, and charging connectors. It is essential to guarantee E.V. owners a smooth experience across platforms and locations.

Environmental Impact

Environmental issues are associated with the production of E.V.s, especially when it comes to exploiting rare earth minerals needed for batteries. Recyclable battery solutions and sustainable mining techniques are essential for reducing this impact.

Urban Planning and Zoning

The creation of EV-ready communities is hampered by zoning laws and urban planning strategies that do not consider electric mobility. To ease the transition, these regulations must be updated and modified.

Consumer Adoption

Several variables, including price, driving distance, and charging infrastructure, affect how widely consumers embrace electric vehicles. A mix of financial incentives, education, and better E.V. products is needed to overcome these obstacles.

Public Perception

There may be a barrier in the form of public perception and awareness of electric automobiles. Promoting the advantages of E.V.s and busting myths and misconceptions about them is crucial to establishing acceptance.

Energy Supply

As the number of electric vehicles rises, so will the demand for the distribution and supply of electricity. The energy industry must change to reflect the shifting environment to guarantee a sustainable transition.

Electric Vehicles and Civil Engineering in the Future

With a focus on cleaner, more efficient transportation, innovation, sustainability, and electric vehicle technology, civil engineering and electric vehicles have bright futures. We welcome the possibility of advancement and change as we overcome obstacles.

Intelligent Infrastructure

Given the rise of electric vehicles, civil engineering has an innately smart future. Infrastructure integration will facilitate communication among automobiles, charging stations, traffic control systems, and urban planning instruments. Improved traffic management, real-time data analysis, and dynamic routing are all promised by this interconnection.

Sustainable Materials

Using environmentally friendly building materials, green building techniques, and sustainable design concepts, civil engineering will continue to place a high priority on sustainability. Long-term cost-effectiveness and environmental responsibility are two benefits of this strategy.

Energy Efficiency

More energy efficiency will result from using electric vehicles in civil engineering projects. Energy consumption and resource conservation will be considered throughout roadways, bridges, tunnels, and charging infrastructure design.

EV-Friendly Urban Development

Bike lanes, pedestrian-friendly designs, electric car charging infrastructure, and sustainable transportation options will all be features of future urban development. The goal is to establish urban settings that are clean, efficient, and easily accessible for electric mobility.

Improved Battery Technology

The development of batteries will determine how electric cars operate in the future. We may anticipate more energy-dense, long-lasting, and sustainable batteries due to research into new materials, lithium-air batteries, and solid-state batteries.

Zero-Emission Transportation

This idea of transportation with no emissions is getting closer. The transportation sector is expected to notably decrease greenhouse gas emissions due to the growing popularity of electric vehicles and renewable energy sources.

Electrified Public Transportation

To reduce noise, air pollution, and fuel expenses, public transportation providers will keep electrifying their fleets. They help to create cleaner, more livable urban areas by doing this.

Intermodal Transportation

Hubs for multimodal transportation, where electric cars and high-speed rail meet, will be essential to mobility in the future. These hubs promote environmentally friendly multimodal transit while providing smooth travel experiences.

Environmental Resilience

Applying resilience concepts to civil engineering projects would improve infrastructure's resistance to natural disasters, guaranteeing that transportation will continue to function dependably in trying circumstances.

Global Cooperation

The development of civil engineering projects and transportation electrification are international undertakings. Cooperation between governments, businesses, and nations will spur growth and guarantee that electric transportation becomes a reality everywhere.

To sum up, the junction of civil engineering and electric vehicles presents problems and opportunities for the future. This journey will be characterized by innovation, sustainability, and the never-ending quest for cleaner, more efficient modes of transportation. We create the conditions for a future in which electric mobility is not only a reality but also a significant factor in transforming our infrastructure, cities, and environment by tackling obstacles and seizing opportunities. We will delve deeper into the particular difficulties and potential opportunities in the upcoming parts to give you a clear picture of the road ahead (Digalwar et al. 2023).

Conclusion

Now that we have completed our investigation into the complex relationship between electric vehicles (E.V.s) and civil engineering, it is appropriate to consider the main conclusions drawn from our work, recognize the continuous advancement of E.V. infrastructure and technology, and look ahead to a time when sustainability, innovation, and cleaner mobility will continue to be the driving forces behind our progress.Over this extensive voyage, we have learned a great deal about the symbiotic relationship between civil engineering and electric car technologies. These important discoveries, which serve as the foundation for the rapid development of our transportation environment, must be summed up and highlighted.

Sustainability at the Core

Including electric vehicles in our transportation system is based on sustainability. As part of the worldwide effort to battle climate change, electric mobility promises to reduce greenhouse gas emissions and improve air quality considerably.

Interconnected Infrastructure

There will be close ties between electric vehicles and civil engineering in the future. Intelligent, networked infrastructure development offers real-time data analysis, dynamic routing, and EV-friendly, energy-efficient urban settings.

Battery Technology Advancement

The success of electric vehicles largely depends on the development of battery technology. Future developments in solid-state batteries, lithium-air batteries, and sustainable materials could result in more energy-dense, durable, and ecologically friendly batteries.

Expansion of the Charging Infrastructure

Widespread E.V. adoption depends on a robust and convenient charging infrastructure. To create charging networks that are easily accessible, standardized, and scalable for both long-distance and urban travelers, the public and commercial sectors must work together.

Urban Design for Electric Mobility

Future urban design is essentially EV-friendly. Bike lanes, pedestrian-friendly designs, infrastructure for charging electric vehicles, and environmentally friendly transportation choices are given top priority. The goal is to establish urban settings that are clean, efficient, and easily accessible for electric mobility.

Electrification of Public Transportation

Around the world, public transportation agencies are electrifying their fleets to reduce noise, air pollution, and fuel expenses. By doing this, they provide easy, sustainable transit options and help to create cleaner, more livable metropolitan areas.

High-Speed Rail Integration

Electric vehicles and high-speed rail are potent combos. Rapid long-distance travel between large cities is made possible by high-speed rail, while flexible and environmentally friendly last-mile connectivity is provided by electric vehicles. When combined, they lessen pollution and traffic.

Global Cooperation

Electrifying transportation is an international project. Together, governments, businesses, and nations can advance the cause and make electric mobility a reality everywhere.

Adaptive Regulation

To promote innovation, infrastructure development, and the adoption of electric vehicles, regulatory frameworks that are flexible enough to allow for electric mobility are crucial. Laws must adapt to the shifting demands of the electric vehicle market.

Environmental Resilience

Adding resilience concepts to civil engineering projects makes infrastructure more resilient to natural disasters, guaranteeing that transportation will function dependably even in the most trying circumstances.

The Ongoing Evolution of E.V. Technology and Infrastructure

The development of electric cars and civil engineering is a never-ending process, constantly changing to meet the demands and overcome obstacles of the modern world. It is critical to acknowledge the changing nature of E.V. infrastructure and technology and the exciting opportunities ahead as we wrap up our investigation.

Innovation and Research

Ongoing innovation and research are inextricably linked to the development of E.V. technology. The future of electric mobility will continue to be shaped by innovative breakthroughs in battery technology, lightweight materials, and energy-efficient charging options.

Grid Integration

Future infrastructure development will center on integrating electric vehicles charging with the electrical grid. Increased reliance on renewable energy sources, lower energy costs, and better grid balancing are all promised by this integration.

Vehicle-to-Grid Technology

E.V.s can send extra energy back into the grid thanks to vehicle-to-grid (V2G) technology, which is expected to play a big part in the future. With the help of this technology, grid demand may be balanced, integration of renewable energy sources may be supported, and E.V. owners' energy expenses may be decreased.

Transportation and Smart Cities

As these cities develop, they prioritize data-driven transportation management, networked infrastructure, and electrified urban design. These advancements offer improved urban living conditions in addition to adequate mobility.

Infrastructure Resilience

Infrastructure projects will give greater weight to climate change and natural disaster resilience. Civil engineering will integrate sustainability and adaptation into its fundamental principles to design infrastructure that can endure environmental difficulties.

Global Electrification

As countries commit to greener, more sustainable mobility, transportation electrification will progress globally. International collaboration and standardized procedures will be essential for building a smooth global transportation network as more nations adopt E.V.s.

Inclusivity and Accessibility

In the future, efforts will be directed at guaranteeing that electric transport is accessible and inclusive to all. This entails attending to the requirements of marginalized populations, lowering the cost of electric cars, and setting up infrastructure for charging them in formerly underdeveloped areas.

Public Awareness

Education and public awareness campaigns about electric vehicles will continue to be crucial. It will be essential to debunk myths, highlight the advantages of adopting E.V.s, and support educated consumer decision-making to promote acceptance and transition.

Electrification of Commercial Transport

To cut emissions and promote sustainability in the logistics industry, there will be a growing push for the electrification of commercial transportation, particularly freight and delivery trucks.

Economic Growth

Projects involving civil engineering and transportation electrification will keep the economy growing. Communities and nations will benefit from innovation, job creation, and sustainable development.

In conclusion, developing electric cars and civil engineering is a continual process leading to cleaner, more effective, and sustainable mobility rather than a definitive destination. Our activities are guided by lessons learned, we innovate to overcome obstacles, and the future looks to change the transportation scene. Rather than being a theoretical concept, we embrace the idea of a future in which electric mobility catalyzes the transition to a more sustainable and interconnected society.

Appendix

It's essential to understand the vocabulary and have access to additional resources and references when it comes to civil engineering and electric cars (E.V.s). The language of electric mobility can be better understood with the help of this appendix. It offers a list of well-chosen resources and a glossary of terms for anybody interested in learning more about this quickly developing field.

Glossary of Terms

To navigate the intricate world of electric cars and civil engineering, one must know the technical terms to characterize this quickly evolving sector. The following glossary provides brief definitions for significant words and expressions:

- A plug-in hybrid electric vehicle (PHEV), a hybrid electric vehicle (HEV), or a battery-electric vehicle (BEV) are the three most common types of electric cars (E.V.s). An E.V. is any automobile that operates entirely or partially on electricity.
- Charging Infrastructure: This is the network of home chargers, public charging stations, and fast-charging networks used to refuel electric car batteries.
- Range Anxiety: This fear, which often prevents people from converting to electric cars, is the concern that an E.V.'s battery may run out of power before it gets to a charging station.
- The science and engineering of designing and building batteries for electric vehicles, focusing on improving longevity, energy density, and charge-discharge efficiency, is known as battery technology.
- Charging Time: The amount of time needed to recharge an electric vehicle's battery depends on the charging pace and battery condition.
- High-Speed Rail (HSR): A kind of passenger rail service that offers quick, economical, and eco-friendly travel far faster than traditional trains.
- Last-mile connectivity: The last mile of a passenger's travel to their destination, which is frequently completed by a variety of methods, such as electric cars, following their departure from a public transportation or high-speed train system.
- Solid-State Batteries: This technology promises better energy density, quicker charging times, and enhanced safety by substituting solid conducting elements for liquid or gel electrolytes.
- Lithium-Air Batteries: These experimental battery technologies can potentially have a high energy density and an extended driving range since they oxidize lithium using air oxygen.
- Smart Cities: Urban regions that use data, technology, and infrastructure to optimize a range of services, including transportation, and to raise the standard of living for their citizens.
- Public-Private Partnerships (PPP): cooperative agreements between public and private organizations, frequently used to fund, plan, build, and run infrastructure projects.
- Transit-Oriented Development (TOD): A strategy to design that encourages people to rely less on private vehicles by concentrating on building walkable, mixed-use communities around public transportation hubs.

- Vehicle-to-grid (V2G) technology allows electric cars to release energy from their batteries into the electrical grid, supporting the grid and generating revenue for owners of the vehicles.
- Sustainable Materials: Building supplies and methods that prioritize environmental stewardship and resource preservation, resulting in environmentally friendly infrastructure.
- Infrastructure Resilience: The capacity of infrastructure to endure and bounce back from extreme events, like natural catastrophes and the effects of climate change, guaranteeing the continuous operation of transportation networks.

Additional Resources and References

For those eager to delve deeper into the world of electric vehicles and their integration with civil engineering, a wealth of resources and references await. Here, we offer a curated selection of sources, websites, and publications to aid in your exploration:

- *National Renewable Energy Laboratory (NREL):* A prominent research institution focusing on renewable energy and advanced vehicle technologies. Their website offers many reports and resources on E.V. technology and infrastructure.
- *U.S. Department of Energy (DOE)*: Office of Energy Efficiency & Renewable Energy: The DOE provides comprehensive information on E.V.s, charging infrastructure, and the latest advancements in clean transportation.
- *International Energy Agency (IEA):* IEA offers in-depth reports and analysis on electric mobility, including trends, market developments, and policy recommendations.
- *The World Electric Vehicle Association (WEVA):* WEVA is a global platform for information and collaboration in the electric vehicle industry. Their website hosts a trove of research and reports on electric mobility.
- *The Smart Electric Power Alliance (SEPA):* SEPA aims to combine electric vehicles with renewable energy. Their writings investigate the relationship between the grid and electric mobility.

- *Transportation Research Board (TRB):* A division of the United States National Academies, TRB provides a wide range of publications and research papers on transportation, including sustainable and electric mobility.
- *EV-Ready Cities and Smart Cities efforts:* Reports and websites from several cities, including Singapore, Oslo, Amsterdam, Barcelona, and Amsterdam, describe their EV-ready city and smart city projects and offer examples of successful efforts from the real world.
- *Scientific Journals:* Scholarly works on many facets of electric mobility and infrastructure may be found in scientific journals, including "Transportation Research Part D: Transport and Environment," "Journal of Power Sources," and "Electric Vehicle Infrastructure."
- *Civil Engineering and Infrastructure Magazines:* Periodicals such as "Infrastructure News" and "Civil Engineering Magazine" frequently publish articles about recent developments in electric vehicle-related civil engineering and infrastructure projects.
- *Academic Institutions:* Research papers and publications about electric cars and their integration with civil engineering are often found on the websites of universities and research centers. Reputable organizations with divisions focused on energy and transportation are excellent resources.
- *Official Websites:* Websites for state and federal governments, as well as the U.S. Department of Transportation, include reports, policy documents, and project updates about the development of E.V. infrastructure.
- *Environmental Organizations:* Resources and studies on electric cars, sustainability, and transportation electrification are available from groups like the Environmental Defense Fund and the Union of Concerned Scientists.

References

Aman Jain, Nipurn Agrawal, Vinay Agrawal, Electric Vehicle Adoption in Rajasthan: Trends, Challenges and Strategies, *IEEE Xplore*, 2023, 1-5.

Digalwar, Abhijeet K., Arpit Rastogi, Assessments of social factors responsible for adoption of electric vehicles in India: a case study, *International Journal of Energy Sector Management*, 17(2), 2023.

Dinh Van Hiep, Nam Hoai Tran, Nguyen Anh Tuan, Tran Manh Hung, Ngo Viet Duc, Hoang Tung, Assessment of Electric Two-Wheelers Development in Establishing a National E-Mobility Roadmap to Promote Sustainable Transport in Vietnam, *Sustainability* 2023, 15(9), 7411.

Furkan Karakuş, Alper Çiçek, Ozan Erdinç, Integration of electric vehicle parking lots into railway network considering line losses: A case study of Istanbul M1 metro line, *Journal of Energy Storage*, 63, 2023, 107101.

Kunpeng Li, Lan Wang, Optimal electric vehicle subsidy and pricing decisions with consideration of EV anxiety and EV preference in green and non-green consumers, *Transportation Research Part E*, 170, 2023, 103010.

Ligao Bao, Motoi Kusadokoro, Atsushi Chitose, Chuangbin Chen, *Travel Behaviour and Society*, 30, 2023, 60-73.

Martins, H., C. O. Henriques, J. R. Figueira, C. S. Silva, A. S. Costa, Assessing policy interventions to stimulate the transition of electric vehicle technology in the European Union, *Socio-Economic Planning Sciences*, 87, Part B, 2023, 101505.

Mohammad Mohammadi, Jesse Thornburg, and Javad Mohammadi, Towards an Energy Future with Ubiquitous Electric Vehicles: Barriers and Opportunities, *Energies 2023*, 16(17), 6379.

Munir Ahmad, Elma Satrovic, How do transportation-based environmental taxation and globalization contribute to ecological sustainability?, *Ecological Informatics*, 74, 2023, 102009.

Paulina Golinska-Dawson, Kanchana Sethanan, Sustainable Urban Freight for Energy-Efficient Smart Cities—Systematic Literature Review, *Energies 2023*, 16(6), 2617.

Priyadarshan Patil, Khashayar Kazemzadeh, Prateek Bansal, *Sustainable Cities and Society*, 88, 2023, 104265.

Saswati Sarmah, Lakhanlal, Biraj Kumar Kakati, Dhanapati Deka, Recent advancement in rechargeable battery technologies, *WIREs Energy and Environment*, 12(2), 2023, e461.

Tapan Senapati, Vladimir Simic, Abhijit Saha, MomciloDobrodolac, Yuan Rong, Erfan Babaee Tirkolaee, Intuitionistic fuzzy power Aczel-Alsina model for prioritization of sustainable transportation sharing practices, *Engineering Applications of Artificial Intelligence*, 119, 2023, 105716.

Vasco Silva, António Amaral, Tânia Fontes, *Sustainable* Urban Last-Mile Logistics: A Systematic Literature Review, *Sustainability 2023*, 15(3), 2285.

Wei Liu, Xin Li, Chunyan Liu, Minxi Wang, Litao Liu, Resilience assessment of the cobalt supply chain in China under the impact of electric vehicles and geopolitical supply risks, *Resources Policy, 80, 2023*, 103183.

Chapter 5

Battery Management

Mohit Hemath Kumar[1,*]
Sourabh Mandol[1]
B. M. Girish[1]
V. V. Vamsi Krishna[2]
Vasavi Annapureddy[3]
and Rajesh Kumar[1]

[1]Department of Mechanical Engineering, Alliance College of Engineering and Design, Alliance University, Bengaluru, Karnataka, India
[2]Department of Mechanical Engineering, JNTU Ananthapur, Andhra Pradesh, India
[3]Department of Computer Science and Engineering, Alliance College of Engineering and Design, Alliance University, Bengaluru, Karnataka, India

Abstract

In summary, batteries are fundamentally responsible for the energy that fuels everything in our modern world, including electric cars reshaping transportation and our everyday electronics. The complexities of battery chemistry and its essential function in energy storage are examined in this abstract. As the workhorses of modern gadgets, lithium-ion batteries have transformed our digital lives. However, many more developing battery technologies promise improved performance and a more environmentally friendly future; thus, the path of battery technology goes beyond lithium-ion. Several essential elements, including sustainability, conductive bridge electrolytes, safety concerns, energy density, and battery management systems, support the development of batteries. These factors guarantee the environmental impact of batteries in addition

---

[*] Corresponding Author's Email: mohitnano1990@gmail.com.

In: Electric Vehicle Technology Structure, Instrumentation and Challenges
Editors: S. N. Sundaram, P. N. Sundaram, M. H. Kumar et al.
ISBN: 979-8-89113-695-3

to their dependability. Furthermore, improved batteries provide greener and more effective solutions in various commercial and industrial industries. Batteries' performance depends on several interrelated factors, such as cycle life, cost, and power density. These variables are significant for electric cars, as maximizing Economy and usefulness requires striking a balance between weight, range, and charging speed. Exciting developments in battery chemistry, infrastructure construction, and electric car technology are anticipated; a dedication to safety, scalability, and regulatory compliance will accompany these developments.

Keywords: cycle life, energy density, power safety, regulatory compliance, safety compliance, user-friendly technology

Introduction

Modern society cannot exist without electricity, crucial for several other industries, including transportation and portable gadgets. Batteries are essential for capturing and storing energy. The fundamental idea of these power storage devices is the same, despite their many sizes and forms: the transformation of chemical energy into electrical energy. This thorough investigation explores the field of battery chemistry, which lies at the core of batteries. We can unleash the promise of electric power in an increasingly electrified world by comprehending the complexities of battery chemistry (Zhou et al. 2023; Xiang et al. 2023).

The Chemistry of Batteries at Its Core: Energy Contained in Chemical Processes

Revealing the Chemical Mechanisms of Battery Functions

Batteries are electrochemical devices, meaning they produce electricity through chemical processes. They comprise an electrolyte, two electrodes (the anode and cathode), and a third component. Electrons are transferred from one electrode to another through the external circuit when a battery is linked to a course. This is because chemical processes take place within the battery cell. An electrical current produced by this electron flow can be used for several purposes.

Types of Batteries and Their Chemistry

Many types of batteries are designed for specific uses, making the world of batteries broad and diverse. Among the most popular battery chemistries are the following:

- *Lithium-Ion (Li-ion) Batteries*: These batteries are widely used in electric cars and portable gadgets due to their high energy density.
- *Lead-Acid:* These batteries are inexpensive and have a high current output; they are frequently utilized in automobile applications.
- *Nickel-Metal Hydride (NiMH):* These semiconductors balance performance and capacity and are used in portable devices and hybrid cars.
- *Alkaline:* Found in many commonplacc home appliances, these batteries are dependable and long-lasting.

Lithium-Ion Batteries: Modern Electronics' Powerhouse

Lithium-Ion Batteries' Ascent

Lithium-ion (Li-ion) batteries are the standard option for electric vehicles (EVs) and have become the mainstay of contemporary portable devices, ranging from laptops to smartphones. The transfer of lithium ions between the anode and cathode, which permits the storage and release of electrical energy, is the central mechanism of their chemistry. Because Li-ion batteries have a high energy density and are incredibly lightweight due to the selection of lithium ions, they are ideal for applications where weight and energy storage capacity are critical.

Lithium-Ion Battery Chemistry

Li-ion batteries use an anode, generally made of graphite, and a cathode, commonly composed of lithium cobalt oxide (LiCoO2) or other compounds based on lithium. Lithium ions travel between the anode and cathode via the electrolyte, frequently a lithium salt in a solvent, during charge and discharge cycles. Lithium ions go from the cathode to the anode during charging and lodge themselves in its structure. These ions return to the cathode during battery discharge, releasing electrical energy.

Investigating Up-and-Coming Battery Technologies: Beyond Lithium-Ion

Solid State Batteries: The Potential for Increased Energy Density and Safety

Even though lithium-ion batteries have advanced significantly, scientists and engineers always look into new battery chemistries. One of the best substitutes is a solid-state battery. These batteries use a solid electrolyte instead of the liquid electrolyte used in Li-ion batteries, which may improve safety, boost energy density, and lengthen cycle life. Solid-state batteries have a lot of promise for electric vehicles and other applications because they lower the danger of leakage and thermal runaway.

Batteries with Lithium-Sulfur: An Advance in Energy Density

Research on lithium-sulfur (Li-S) batteries is also quite active. Sulfur is the cathode material in these batteries, producing a greater energy density than conventional lithium-ion batteries. Sulfur is plentiful and lightweight, which makes Li-S batteries a desirable option for electric cars. Even though there are still difficulties, such as how to dissolve lithium polysulfides, research is still being done to realize this chemical's promise fully.

The Pursuit of Energy Density: The Engine Powering Battery Development

Definition of Energy Density

Energy density, which expresses how much energy can be stored in a battery within a specific space or weight, is crucial in battery chemistry. High energy density batteries are perfect for uses where importance and space are critical, such as electric cars and portable gadgets, since they can store more energy in a small, light form.

The Effect of Energy Density on Electric Vehicles

Energy density is crucial when it comes to electric cars. It directly impacts the weight and driving range of an EV. High-energy-density batteries enable electric vehicles to go farther between charges, which lowers the frequency of recharging and increases the convenience of using an electric car. However, higher energy density is still being sought as battery research and development are still motivated by this goal.

Battery Chemistry Safety Considerations

Safety Issues and Thermal Runaway

Electric power is made possible by battery chemistry. Still, this process is not without difficulties, especially regarding safety. Thermal runaway, an uncommon but hazardous occurrence where the battery overheats and might result in fires or explosions, is one of the most urgent safety issues. Researchers and manufacturers devote much time and resources to developing better battery chemistries and adding safety measures to reduce these dangers and make electric cars and portable electronics safer for users.

Rendering

The dynamic field of battery chemistry is where invention and science meet to fuel our contemporary lifestyles. Because of their excellent energy density and lightweight construction, lithium-ion batteries have entirely changed the market for electric cars and portable gadgets. However, the adventure is far from over. Future energy storage technologies, such as solid-state and lithium-sulfur batteries, have the potential to provide even safer and more effective options. We are opening the door to a more electrified and sustainable future where electric power will continue to propel advancement and light our lives as we explore further into the complex realm of battery chemistry (Shi et al. 2023; Kim et al. 2023).

Conductive Bridge: Electrolytes

Lithium-ion Battery Liquid Electrolytes

In conventional lithium-ion batteries, the liquid electrolyte connects the anode and cathode. It permits the transfer of lithium ions between the electrodes during the charge and discharge cycles. Liquid electrolytes have been a blessing in disguise, but they have safety issues, most notably the possibility of leaks and thermal runaway.

Sturdy Electrolytes: The Secret to Battery Safety

Solid electrolytes are a viable substitute in the field of battery chemistry. While providing conductivity, these solid-state electrolytes do away with the hazards of liquid electrolytes. These electrolytes can improve safety in solid-state batteries by stopping leaks and lowering the possibility of short circuits, making them a desirable option for various uses.

Battery Management Systems (BMS): Guaranteeing Durability And Security

Overview of BMS

The Battery Management System is an essential component of battery technology (BMS). The battery's condition, including its temperature, charge, and discharge rates, and general health, is monitored by this system. It is essential for extending battery life, preserving security, and enhancing functionality.

Electric Vehicle BMS

BMS is necessary in electric cars to guarantee the battery pack's lifespan. Distributing the workload across the cells equitably helps avoid overcharging, over-discharging, and overheating. The battery management system (BMS) is essential in maintaining acceptable temperature ranges for the battery (Ahasan et al. 2023).

Recycling and Sustainability: Battery Second Life

The Battery Production's Environmental Footprint

Transportation, energy-intensive manufacturing procedures, and raw material extraction are all necessary to create batteries, particularly lithium-ion batteries. These elements add to the battery production process's environmental impact. Sustainable battery chemistry considers the battery's complete lifespan, from extraction of raw materials to disposal.

Battery Recycling: An Increasing Need

An essential component of sustainable battery chemistry is recycling. End-of-life batteries are gathered and reprocessed to recover valuable materials and minimize waste. Metals that can be used again to create new batteries, such as nickel, cobalt, and lithium, can be extracted through the recycling process. The increasing popularity of electric cars necessitates the development of an adequate recycling infrastructure to reduce the environmental impact of battery disposal.

Advanced Batteries: Commercial and Industrial Uses

Energy Storage at the Utility Scale

Utility-scale energy storage has found use for cutting-edge battery technology. These large-scale systems may store surplus renewable energy, including solar or wind power, during peak demand or when renewable sources aren't producing electricity. They frequently employ Li-ion or developing chemistries for this purpose. Utility-scale energy storage lessens the fossil fuel-based backup power requirement while also assisting in grid stabilization and increased energy dependability.

Aviation and Aerospace

The aircraft and aerospace sectors are transforming because of battery chemistry. Lithium-ion batteries increasingly power auxiliary power units and emergency power systems in airplanes. Electric aircraft promises lower emissions and operating costs, and the aviation industry is adopting electric propulsion due to the development of lighter and higher-capacity batteries.

Upcoming Aspects of Battery Chemistry

A Look into the Future with Quantum Batteries

One of the most innovative areas of battery chemistry research is quantum batteries. These batteries, which use quantum physics as their foundation, have the potential to completely transform energy storage thanks to their incredibly high energy density and quick charging times. If successful, quantum batteries may significantly increase the driving range of electric cars and open up new uses across several sectors.

Battery Chemistry: Environmental Considerations

Environmental impact and sustainability are critical to the development of battery chemistry. Researchers are looking into more eco-friendly techniques and materials. The manufacture and disposal of batteries will have less carbon impact in the future because of innovations like recyclable and biodegradable battery components.

Rendering

The field of battery chemistry is dynamic, constantly changing and pushing the envelope of what is conceivable. The physics of batteries continues to spur innovation across sectors, from solid-state batteries that promise safer energy storage to lithium-ion batteries that have become indispensable to everyday life. The development of quantum batteries, infrastructure for recycling, and solid-state electrolyte improvements will usher in a new era of energy storage that will fundamentally alter our planet. With batteries acting as a catalyst for advancement, creativity, and environmental responsibility, the potential of a more electrified and sustainable future becomes more apparent as we traverse this always-changing field of battery chemistry.

Power Density: The Key to Outstanding Efficiency

The Power Density Revealed

Power density is concerned with the quick distribution of stored energy, whereas energy density is more concerned with storage. Fast energy delivery from high-power density batteries is essential for quick acceleration, agile handling, and an all-around thrilling driving experience. Power density is an essential metric in battery chemistry because it controls how rapidly the battery can release energy. It gauges how quickly the battery can deliver power, which is critical information for applications like electric automobiles that need to use small amounts of energy often.

EV Experience Elevation

Power density is critical to the quick torque and smooth acceleration for which electric vehicles are praised. Choosing a high-power density battery chemistry guarantees that an electric car will continue operating as intended, completely changing how we think about electric vehicles. Manufacturers of electric vehicles are constantly working to improve the batteries' power density so that the driving experience can compete with that of cars with conventional internal combustion engines. Improvements in power density make electric vehicles more competitive and consumer-friendly by enabling faster acceleration, better handling, and responsiveness.

Cycle Life: The Length and Dependability Metric

Comprehending Life Cycles

When assessing batteries, cycle life is an important parameter. It speaks to how many cycles of charge and discharge a battery can withstand before seeing a noticeable reduction in capacity. Stated differently, it assesses how long a battery can last and how well it can continue to function over time. A battery with a more excellent cycle life is likely more dependable and long-lasting. It indicates how many cycles of charging and discharging the battery can endure before there is a discernible loss in capacity.

Consequences for the Economy

Cycle life is crucial to an electric car owner's Total Cost of Ownership (TCO). Longer cycle life batteries require fewer replacements, which lowers maintenance costs and improves the economic feasibility of electric cars. The economics and sustainability of electric transportation are inextricably related to the search for batteries that persist and outlive their predecessors. When evaluating an electric vehicle's long-term cost-effectiveness, cycle life is an important consideration. A battery with a longer cycle life needs fewer replacements throughout the vehicle's life, which lowers the total cost of ownership. Research and development on batteries is continuously focused on finding ways to increase their cycle life. Researchers and producers aim to create designs and chemistries that are resistant to severe deterioration over an extended period of use. This lessens the environmental effect of battery manufacture and disposal, which helps owners of electric vehicles and advances the sustainability of electric transportation.

Cost: Juggling Affordability and Efficiency

The Complexities of Expense Factors

Beyond the initial purchase price, other factors contribute to the overall cost of a battery. Total Cost of Ownership (TCO) review thoroughly considers operational expenses, purchase costs, and possible resale value.

Economies of Scale and Mass Production

The economics of battery manufacture are closely related to the affordability of electric cars. Batteries always get cheaper as production quantities rise and economies of scale take effect. Electric vehicles are becoming more affordable for a broader range of users, which advances the cause of clean and sustainable transportation. Economies of scale significantly influence the price of batteries for electric vehicles. The cost of the battery per unit drops as production volume rises. This decrease is credited to better production procedures, more automation, and the capacity to bargain for lower raw material costs. Technological developments are also fueled by mass production, which results in more economical and efficient battery manufacturing processes. The industry gains from these cost savings as the market for electric cars expands, boosting the competitiveness of electric vehicles relative to their internal combustion engine competitors.

Power density, cycle life, and cost are essential factors in battery chemistry. The performance of an electric vehicle is determined by its power density, which affects its acceleration and reactivity. Power density enhancement plays a crucial role in increasing the attractiveness and competitiveness of electric cars. The Total Cost of Ownership (TCO) of electric cars is directly impacted by cycle life, as longer-lasting batteries lower maintenance costs and increase the economic feasibility of electric mobility. Battery research aims to build more dependable and durable batteries by extending their cycle life. The cost of an electric vehicle's initial purchase and its total cost of ownership over time are both taken into account. Batteries are becoming more economical and available for various consumers because of economies of scale and mass manufacture. The quest for increased power density, cycle life, and cost-effectiveness is crucial to improving the sustainability and affordability of electric mobility as the electric car market grows.

The Effect of Battery Energy Density on Power Density

Energy Density and Power Density's Relationship

In battery technology, energy density and power density are closely related. As previously said, energy density is the quantity of energy a battery can hold in relation to its volume or weight. However, power density indicates the rate at which a battery may release its stored energy. The performance of electric cars is greatly influenced by the interaction between these two characteristics, which are not mutually exclusive and frequently enhance one another. The ability to store a substantial quantity of energy in a small area is made possible by high energy density batteries, which extend the driving range of electric cars. Power density is critical because, when the vehicle accelerates, this energy needs to be transmitted as effectively as possible. High-power-density batteries can deliver the short bursts of power required for responsive handling and rapid acceleration.

Progress in Power Density and Energy

Battery chemistry research and development goals are still to increase power and energy density. Breakthroughs aid cell architecture, electrode design, and materials science developments. The aim is to achieve a balance that allows electric vehicles to provide longer driving ranges without sacrificing outstanding performance. Energy and power density advancements significantly impact electric vehicle design and consumer adoption. Electric cars with a higher energy density have a more extended range, which lessens range anxiety and increases their suitability for daily use. In addition, higher power density makes electric cars more competitive with conventional internal combustion engine vehicles by improving acceleration and the entire driving experience.

Electric Vehicle Performance and Power Density

Power Density's Function in Electric Car Acceleration

Electric cars stand out for their quick acceleration and instantaneous torque delivery. The excellent power density of batteries used in electric vehicles is responsible for this feature. How rapidly the battery can release energy to power the electric motor and move the car ahead is determined by its power density. High-power density battery electric cars can remarkably accelerate, frequently surpassing their internal combustion engine competitors. This feature not only improves the overall performance of electric vehicles but also makes driving them more exciting. Power density is essential for electric cars to be competitive in terms of speed, responsiveness, and handling.

Implications for the Design of Electric Vehicles

Manufacturers of electric cars strive to achieve optimal power density to guarantee superior performance from their vehicles. It takes careful engineering of the battery pack, electric motor, and control systems to achieve excellent power density. These parts must cooperate for electric cars to deliver the quick acceleration and responsiveness that buyers want. The development of electric cars includes higher power density, energy management techniques, and regenerative braking technologies. By recovering and storing energy during deceleration and using it for acceleration, these technologies let electric cars achieve even higher power densities and overall efficiency.

Cycle Life: The Secret To Sustainability and Longevity

Battery Chemistry's Cycle Life

As was previously said, cycle life is the number of charge-discharge cycles a battery can withstand before seeing a significant decline in capacity. Given that it directly affects a battery's longevity and dependability, it is an essential metric for consumer and industrial applications. In addition to lowering the need for frequent battery changes, a longer cycle life promotes sustainability by lessening the environmental effect of battery manufacture and disposal. Research and development efforts primarily focus on extending cycle life to produce more dependable and long-lasting batteries.

Consequences for Electric Cars

Cycle life is crucial when it comes to electric cars. Owners anticipate thousands of charge-discharge cycles from their batteries throughout an electric vehicle's lifespan. Longer cycle life batteries lower maintenance costs and increase the economic feasibility of electric cars. Managing operating conditions, choosing suitable material, and designing electrodes may achieve long cycle life. Scientists constantly enhance these features to produce batteries that sustain heavy cycling without losing functionality.

Expense: Judging between Cost and Value

The Complicated Cost Formula

Beyond the initial purchase price, other factors contribute to the overall cost of batteries. Some elements are considered while assessing the Total Cost of Ownership (TCO) of electric cars, including purchase costs, running expenses, and possible resale value.

Economies of Scale and Mass Production

The economics of battery manufacture are closely related to the affordability of electric cars. Batteries always get cheaper as production quantities rise and economies of scale take effect. Electric vehicles are becoming more affordable for a broader range of users, which advances the cause of clean and sustainable transportation. Economies of scale significantly influence the price of batteries for electric vehicles. The cost of the battery per unit drops as production volume rises. This decrease is credited to better production procedures, more automation, and the capacity to bargain for lower raw material costs. Technological developments are also fueled by mass production, which results in more economical and efficient battery manufacturing processes. The industry gains from these cost savings as the market for electric cars expands, boosting the competitiveness of electric vehicles relative to their internal combustion engine competitors. Regarding battery chemistry, power density, cycle life, and cost are crucial, particularly considering how they will affect electric cars. Power density guarantees outstanding performance and acceleration, increasing electric vehicles' competitiveness and customer appeal. Improvements in power density lead to longer driving ranges and better driving experiences when paired with high energy density. Cycle life is a crucial indicator of a battery's dependability and sustainability. Longer-lasting batteries minimize the environmental effect of battery disposal by lowering maintenance expenses and the need for replacements. Researchers are still concentrating on increasing cycle life to increase the robustness and reliability of batteries. Electric car Cost concerns include the initial purchase price and the overall Total Cost of Ownership (TCO). Batteries are becoming more economical and available for various consumers because of economies of scale and mass manufacture. The quest for increased power density, cycle life, and cost-effectiveness is crucial to improving the sustainability and affordability of electric mobility as the electric car market grows (Zhang et al. 2023; Lv et al. 2023).

Charging Velocity: Increasing Electric Vehicle Progression

How Faster Charging Helps Electric Cars

One essential feature of electric cars (EVs) that directly affects their convenience and utility is charging speed. It describes the rate at which an electric car's battery can be charged, enabling drivers to restore their energy and resume driving swiftly. Quick and effective charging is crucial to increase the practicality and attraction of electric mobility to a wider spectrum of customers.

Charging Types

Electric car charging is accessible at different speeds in public locations and at home at different levels. Among these options for charging are:

Level 1 Charging
This is the slowest level of charging using a regular home plug. Its low power output makes it ideal for charging overnight at home.

Level 2 Charging
Installable at home or public charging stations, Level 2 chargers are more potent. They are a well-liked option for home charging as they offer a quicker charging speed.

DC Fast Charging
This is the fastest way to charge electric vehicles. With its ability to quickly deliver a substantial quantity of energy, it is perfect for both rapid top-ups and long-distance trips.

Range of Electric Vehicles and Charging Speed

The pace at which an electric car charges directly correlates to its range. Quicker replenishment of an electric car reduces downtime and enables drivers to go farther in a single day thanks to faster charging. The broad adoption of electric vehicles is contingent upon an efficient charging infrastructure. The charging speed of electric cars is increasing dramatically as charging technology advances, including the creation of high-power DC fast charging networks.

Range: Getting Past the Fear of the Unknown

Electric Vehicle Range Definition

An electric vehicle's range is the maximum distance between charges. It is a crucial component for customers as it establishes whether an electric car is useful and practicable for their requirements. The worry of running out of power before arriving at a charging station, or range anxiety, has been a significant barrier to adopting electric vehicles.

Range-Influential Factors

An electric vehicle's range is influenced by some factors, including:

Battery Energy Density

The battery's energy density determines how much energy it can hold and, in turn, how far the car can go between charges.

Driving Conditions

The efficiency and range of the vehicle might be impacted by driving at more incredible speeds, in extremely hot or cold climates, or on steep terrain.

Aerodynamics

The vehicle's design, in particular its aerodynamics, can affect range and, consequently, energy consumption.

Driver Behavior

Regenerative braking and gentle driving can increase range, whereas aggressive and fast acceleration can reduce it.

Extension of the Range of Electric Vehicles

The range of electric vehicles is becoming longer as energy density and battery technologies advance. These days, a single charge can propel modern electric cars well over 200 miles, and some top-of-the-line versions may go up to 300 miles. Consumers place a high value on the range, and automakers are making concerted efforts to decrease range concerns. The usefulness of electric cars is further improved by the development of quicker charging technologies and the availability of more extensive charging networks.

Weight: Juggling Efficiency and Performance

Weighing Issues for Electric Cars

For electric cars to operate well and efficiently, weight is essential. It has an impact on handling, acceleration, and total energy usage. Electric vehicle makers strive to balance weight, performance, and efficiency to build cars that satisfy consumer demands.

Weight of Battery

The battery pack of an electric vehicle is the primary source of its weight. The total weight, handling, and Economy of the car can all be strongly impacted by the weight of the battery. It is crucial to design batteries with a high energy density while conserving weight since heavier batteries can potentially decrease efficiency and range.

Vehicle Design and Lightweight Materials

Manufacturers are using lightweight materials like carbon fiber, aluminum, and composite materials in electric vehicle designs to reduce weight. These components increase efficiency and lessen the battery pack's weight. Weight control also involves vehicle design. Its aerodynamic design and weight distribution can influence the efficiency and handling of an electric vehicle. Continuous work is being done to optimize vehicle design to increase the overall Economy and performance of electric cars.

How Weight, Range, and Charging Speed Interact

The Equilibrium Partnership

Weight, range, and charging speed are all related to electric cars. The way these elements interact directly impacts how convenient and appealing electric transportation is. Electric cars' range is affected by how quickly they charge since it reduces driver downtime. Long-distance driving with electric vehicles becomes increasingly feasible as high-speed charging networks increase and charging infrastructure improves. Customers place a high value on an electric vehicle's range as it dictates how far they can go between charges. Electric cars now have ranges that satisfy the demands of many drivers because of developments in battery technology and energy density. Extended ranges alleviate concerns about running out of power and qualify electric cars for daily usage. Weight is a two-edged sword in terms of electric cars. Even though lighter cars are usually more efficient, the battery pack accounts for a sizable amount of an electric vehicle's weight. Creating batteries with a high energy density while limiting weight is a problem for manufacturers. Lightweight materials and creative vehicle design are used to achieve this equilibrium.

The Pursuit of Excellence

The technology of electric vehicles is still being refined. The aim is to develop cars with quick charging times, long driving ranges, and balanced weight distribution. Innovations in battery chemistry, charging infrastructure, lightweight materials, and vehicle design are needed to achieve this balance. The development of electric vehicle technology can potentially transform how we travel entirely. We are getting closer to a future where electric transportation is sustainable but also functional, practical, and available to everyone by tackling the interaction of charging speed, range, and weight.

With the combination of these factors, electric cars have a great chance of becoming commonplace transportation.

Speed of Charging: Increasing Electric Mobility

The Significance of Charging Velocity

One of the critical elements affecting the usability and attraction of electric cars (EVs) is charging speed. It describes the rate at which an electric vehicle's battery can be trusted, directly relating to how soon a driver may refuel the car and resume their trip. Fast and efficient charging is becoming increasingly necessary as charging technology develops to increase the convenience and accessibility of electric mobility for a wider variety of users. Charge speed influences the general uptake of electric cars and their usefulness. One major issue with electric mobility is the time it takes to recharge an EV's battery compared to refilling a traditional gasoline car. This may be resolved by having the capacity to fast recharge an EV's battery.

Charging Types and Their Effects

The infrastructure used to charge electric vehicles differs in terms of speed; thus, it's critical to comprehend the various charging levels:

First-Level Charging

This is the most minor fast charging level, usually using a regular home plug. Level 1 charging works best at home for overnight charging because it uses very little power. Long-distance travel is not feasible with it, but everyday commuting is convenient.

Charging at Level Two

More vital Level 2 chargers are available at public charging stations and may be installed at home. They are a well-liked option for home charging as they offer a modest charging pace. Level 2 charging strikes a compromise between convenience and charging time, making it ideal for topping off batteries during shorter pauses.

DC Fast Charging

DC fast charging is the quickest charging method for electric cars. With its ability to deliver significant energy in a brief period, it is perfect for both rapid top-ups and long-distance trips. Range anxiety is reduced by the instantaneous charging speed of DC fast chargers, which also dramatically enhances the allure of electric vehicles.

The Effect of Charging Speed on the Uptake of Electric Vehicles

The adoption of electric vehicles is significantly impacted by charging speed, particularly for people who use their cars for longer trips and everyday commuting. The availability of quick charging infrastructure substantially affects the perceived utility of electric vehicles. The charging speed of electric vehicles is increasing dramatically as charging technology advances, including the creation of high-power DC fast charging networks. Quicker charging speeds lessen the annoyance of charging an electric car and hasten the shift to electric transportation.

Range: The Guarantee of Usability for Electric Vehicles

Range: An Important Factor for Electric Vehicles

Regarding electric vehicles, the range is crucial to Customers, giving the range a lot of thought since it directly impacts how practical an electric car is. Increased range reduces anxiety and solves one of the main issues with adopting electric vehicles. For owners of electric vehicles, the guarantee of having sufficient range to get to a location without having to worry about recharging all the time is essential. Its increased range results from improvements in energy density, battery technology, and overall electric vehicle design.

Variables Affecting the Range of Electric Vehicles

An electric vehicle's range is determined by some factors, all of which must be taken into account when using the car in actual traffic:

Density of Battery

The range of an electric vehicle is primarily dependent on the energy density of the battery. Longer ranges are made possible by batteries with better energy densities, which store more energy in the same volume or mass.

Conditions for Driving

The driving conditions and surroundings may significantly impact the range of an electric vehicle. The car's efficiency can be lowered by driving at faster speeds, in extremely hot or cold weather, or over steep terrain, which will limit range.

Aerodynamics

A vehicle's design, particularly its aerodynamics, significantly influences range and energy consumption. Aerodynamic and streamlined designs contribute to less energy loss, increasing the vehicle's capacity.

Action of the Driver

The range of an electric car can also be affected by how the driver uses it. Frequent high-speed and aggressive driving with quick acceleration can lower the range and increase energy consumption. On the other hand, capacity may be increased via energy-efficient driving, regenerative braking, and smooth driving.

Progress in the Range of Electric Vehicles

The electric car range keeps growing as energy density and battery technology advance. These days, a single charge can propel modern electric cars well over 200 miles, and some top-of-the-line versions may go up to 300 miles. Improvements in range significantly increase electric vehicles' usefulness while reducing range anxiety. As the range of electric cars develops, more customers find electric mobility adequate for their daily demands, including long-distance travel (An et al. 2023).

Future Developments and Infrastructure

Extension of the Charging Infrastructure

For electric vehicles to be widely used, the infrastructure for charging them must be developed and expanded. Governments, private businesses, and public organizations are investing in installing charging stations to make charging more practical and available. DC fast-charging networks are becoming increasingly common, typically found beside significant thoroughfares and roads. The time needed for long-distance driving is significantly decreased by these fast-charging stations, which also help to increase the popularity of electric cars for road trips.

Technology for Wireless Charging

Inductive charging, another name for wireless charging technology, is another cutting-edge advancement that can completely change how electric vehicles are charged. EVs may be charged wirelessly by just parking over a wireless charging pad or coil rather than putting in a charging cable. With the removal of physical connectors, this technology provides a smooth and practical charging experience, increasing accessibility to charging, particularly for urban and commercial applications.

Advances in Battery Technology

Battery technology must continue to evolve to improve charging speed, range, and weight. Researchers are experimenting with new materials and chemistries to increase energy density, shorten charging periods, and improve battery performance overall. For instance, research is being done on lithium-sulfur (Li-S) batteries because of their potential to boost significantly energy density and expand electric vehicles' driving range. In addition, research is being done on solid-state batteries because they may provide excellent safety, a longer cycle life, and a better energy density than liquid electrolytes.

Recycling and Sustainability

Concerns about the sustainability of batteries used in electric vehicles are developing. As more people choose electric cars, there is a greater demand for environmentally friendly methods of producing and recycling batteries. Recycling technology innovations seek to minimize waste and their negative effects on the Environment by recovering valuable materials from end-of-life batteries. Researchers are also looking at more ecologically friendly materials and production techniques to lessen the carbon footprint associated with battery manufacture and disposal. Weight, range, and charging speed are all crucial factors in the continuous development of electric mobility. The interaction of these elements is changing how we think about electric cars and propelling the revolution in transportation. Electric vehicles are becoming more and more competitive with gasoline-powered automobiles as charging technology develops. Various consumers find electric vehicles more enticing due to their longer ranges, which reduce range anxiety. Weight issues are considered through lightweight materials and creative design to improve performance and efficiency.

Electric Vehicle Technology Safety

The Safety of Electric Vehicles Is Crucial

Safety is the first priority when it comes to the operation and design of electric vehicles (EVs). As with any mode of transportation, it is crucial to ensure everyone on the road, including the occupants, is safe. To reduce hazards and improve the overall safety of their cars, electric vehicle producers invest in various safety measures and technology.

Thermal Management and Battery Safety

An electric vehicle's battery pack is essential, and safety is of the utmost importance. EVs frequently employ lithium-ion batteries, which have a high energy density but can occasionally experience thermal runaway. To combat this, EVs have advanced battery management systems that keep an eye on voltage and temperature in the cells. These systems can take action when abnormalities are found, such as lowering the charging rate or turning off the battery to stop overheating or fires. The battery enclosure's design also incorporates safety precautions to avert any possible fire or thermal runaway. Reducing hazards requires using insulation, crash-resistant buildings, and fire-resistant materials. The high-voltage parts of EVs are made to immediately turn off in the event of a severe accident, lowering the possibility of electric shock or fire.

Structural Safety of Electric Vehicles

An electric vehicle's construction is also made with safety in mind. To protect the battery pack from exterior impacts in the case of a collision, it is commonly mounted inside the vehicle's frame. Furthermore, modern safety features in electric cars include airbags, electronic stability control (ESC), antilock braking systems (ABS), and collision avoidance technology. These features all work together to reduce the risk of collisions and safeguard passengers.

Managing Elevated Voltage

Generally speaking, electric cars run at greater voltages than cars with traditional internal combustion engines. Although this increases their effectiveness, it calls for extra safety precautions. First responders, maintenance, and repair staff are all trained to handle high-voltage components safely. Emergency personnel, including paramedics and firefighters, receive specialized training on managing incidents involving electric vehicles. In the event of an accident, they have the expertise and equipment necessary to turn off high-voltage systems and free up passengers safely.

Standards and Regulations' Role

Standards and regulations are essential for guaranteeing the security of electric cars. Electric vehicle manufacturers must abide by safety standards set by government agencies and organizations. These standards cover Numerous topics, including electrical safety, crashworthiness, battery safety, and managing high-voltage systems. Manufacturers must carry out stringent safety testing and comply with these regulations to guarantee that consumers may safely use their electric vehicles. These standards change as technology develops and frequently call for constant improvement and compliance.

Scalability of Manufacturing for Electric Vehicles

The Difficulty of Manufacturing

In contrast to automobiles powered by internal combustion engines, the production of electric vehicles has distinct manufacturing obstacles. The manufacturing of electric cars necessitates specific equipment, parts, and knowledge. The increasing demand for electric vehicles has made the scalability of production processes a crucial factor.

Production of Batteries

A vital aspect of the development of electric vehicles is the creation of batteries. Facilities for producing batteries must be scalable to satisfy the growing demand for electric cars. Manufacturers spend money on research and development to increase the cost-effectiveness and efficiency of battery production.

Automobile Component

Because electric car assembly involves so many diverse parts and systems, it calls for specific knowledge and tools. Manufacturers face difficulty ensuring that assembly lines are flexible and expandable to accommodate different types of electric vehicles. Scalability is the capacity to create various models, integrate new technologies, and streamline manufacturing procedures.

Scalability of Charging Infrastructure

A growing infrastructure for charging electric vehicles is necessary as the demand for these vehicles expands. The capacity to effectively develop, operate, and maintain charging stations is critical to scalability in charging infrastructure. To keep the viability and convenience of electric vehicles for customers, governments, private enterprises, and organizations make investments in the scalability of charging networks.

Teamwork Initiatives

Governments, organizations, suppliers, and manufacturers work together to overcome the issue of scalability in the production of electric vehicles. Partnerships are created to exchange information, combine resources, and create original solutions. These partnerships aim to hasten the creation of scalable production techniques that can satisfy the rising demand for electric cars.

Essential components of electric vehicle technology are safety, environmental effects, and manufacturing scalability. Modern battery management systems, structural safety measures, and high-voltage component handling procedures are all necessary to ensure the safety of electric cars. These precautions are essential for protecting both emergency personnel and building residents. When fueled by renewable energy sources, electric cars have the potential to lessen their environmental impact by producing zero tailpipe emissions and decreasing greenhouse gas emissions. Their environmental friendliness is further increased by increases in energy efficiency and the sourcing and recycling of sustainable materials. For the production of electric vehicles to keep up with the growing demand, scalability is crucial. To ensure that electric vehicles are feasible and affordable for consumers, manufacturers invest in battery production, vehicle assembly, and charging infrastructure. To effectively handle the potential problems presented by electric vehicle technology and accelerate the shift to more environmentally friendly and practical modes of transportation, cooperation amongst many stakeholders is essential.

Technology of Packaging and Integration for Electric Vehicles

The Importance of Integration and Packaging

The development and production of electric vehicles (EVs) depend heavily on packaging and integration. The arrangement and integration of different systems and components of an electric vehicle (EV) affect the vehicle's overall efficiency, safety, and performance. This multifaceted process includes vehicle architecture, thermal management, and component placement.

Space Efficiency and Compact Design

Space efficiency is a crucial design feature of electric cars, which maximizes the configuration and positioning of essential parts. This covers the electric motor, power electronics, and battery pack. The compact design increases the room for passengers and freight while improving aerodynamics and the vehicle's overall efficiency.

Management of Heat

Effective thermal management is essential to guarantee that electric vehicle components function within their designated temperature limits. Particularly susceptible to temperature changes are batteries. Effective cooling techniques, such as liquid or air cooling, keep batteries at their ideal working temperatures and prolong their lifespan. In cold weather, thermal management systems also contribute to the cabin's heating. By capturing and redistributing waste heat produced during battery operation, these systems can lessen energy-intensive cabin heating requirements.

Safety Points to Remember

Integrity and packaging are essential to electric car safety. Components must be located strategically and protected to reduce the possibility of accidents and guarantee occupant safety. To protect the batteries in the case of a collision, the battery enclosure's design and crashworthiness characteristics are crucial. Furthermore, the integration and protection of components in electric cars are governed by safety standards and laws. To guarantee their cars' safety, manufacturers must abide by specific regulations.

Comparing Cell-to-Vehicle (CTV) with Cell-to-Pack (CTP)

Integration of Cell-to-Pack (CtP)

The process of directly integrating individual battery cells into a larger battery pack without needing an intermediary module is called "cell-to-pack" (CtP) integration. This method improves energy density while lowering battery construction costs and complexity. CtP is exceptionally well suited for electric cars as it can result in battery packs that are lighter and smaller. By doing away with modules and extra parts, CtP integration may streamline production and lower the danger of thermal and electrical impedance. Additionally, it makes better use of the available space inside the battery pack to increase its capacity for energy storage.

Integration of Cell-to-Vehicle (CtV)

The more significant integration of the battery pack into the electric vehicle's design is the primary goal of cell-to-vehicle (CtV) integration. In addition to the battery pack's actual location, the vehicle's procedure must have power electronics, thermal management systems, and battery management systems (BMS). The interaction between the battery pack and the other systems in the car, such as the electric motor, regenerative braking, and energy management, is taken into account via CTV integration. It is essential to maximizing the vehicle's overall effectiveness and performance.

Benefits and Constraints

The ease and possible cost savings of CtP integration make it desirable. It may result in lighter, more compact battery packs, enhancing the vehicle's overall weight distribution. The interaction of the battery with other vehicle components could not be fully addressed by CtP integration, which could restrict optimization options. CtV integration provides a thorough method that considers the complete vehicle's architecture. This method makes more control over the battery's interactions with other systems and the vehicle's energy management possible. It could, however, be more complex and call for a higher degree of engineering proficiency. The precise design objectives of the electric vehicle, manufacturing capabilities, and economic concerns are some of the elements that influence the decision between CtP and CtV integration. Many EV manufacturers combine both strategies to achieve a compromise between ease of use and thorough integration.

Supply Chain Factors in the Production of Electric Vehicles

Complexity of the Supply Chain

The intricate network of suppliers, parts, and materials involved in producing electric vehicles is known as the supply chain. Acquiring specialist components like electric motors, power electronics, battery cells, and different car subsystems is part of it. Because the supply chain is worldwide in scope, coordination, and management are necessary to guarantee a consistent and effective flow of resources.

Difficulties with Battery Supply

The battery is one of the most important parts of electric cars. High-quality battery cell supply is sometimes a production barrier for electric vehicles. For manufacturers, ensuring a steady supply of batteries is crucial as demand for electric cars rises. Automakers and battery manufacturers have partnered to address this issue and guarantee a constant supply of high-performance batteries. The ethical and sustainable procurement of raw materials, such as cobalt and lithium, used in battery manufacture, is also considered when planning a battery supply. Manufacturers strive to establish responsible sourcing methods because they are becoming more aware of their supply chains' ethical and environmental implications.

Dependencies Abroad

Because the supply chain for electric vehicles is global, reliance on foreign suppliers is typical. Materials and components could come from other nations, which complicates matters with trade agreements, tariffs, and logistics. International dependencies may also impact the supply chain's resilience. The flow of essential components might be disrupted by natural catastrophes or geopolitical conflicts, which can impact manufacturing schedules and provide difficulties for manufacturers.

Localization and Scalability

Manufacturers are developing plans to make their supply networks more localized and scalable. The capacity to adjust and grow the supply chain in response to rising demand is called scalability. To lessen reliance on foreign suppliers and increase the robustness of the supply chain, localization seeks to acquire parts and materials from close sources.

Sustainability in the Environment

In the production of electric vehicles, supply chain factors go beyond just acquiring parts. Additionally, manufacturers are paying attention to how sustainable their supply networks are. This entails cutting manufacturing process waste, guaranteeing ethical material procurement, and lowering carbon emissions in logistics.

Crucial elements of electric car technology and production include packaging and integration, supply chain issues, and the comparison of cell-to-pack (CtP) vs. cell-to-vehicle (CtV) integration. The practical arrangement and arrangement of parts within an electric car is referred to as packaging and integration. This ensures safety, temperature control, and space efficiency. Depending on particular design objectives and trade-offs, one might choose between integration for cell-to-pack (CtP) and cell-to-vehicle (CtV). It simplifies battery assembly, while CtV considers the battery pack's more comprehensive integration with the vehicle's design. The supply chain involves many intricate factors, such as maintaining a consistent supply of components, controlling global interdependence, and encouraging environmental responsibility and sustainability. These factors are vital in determining the development, manufacture, and sustainability of electric cars as the industry for these vehicles grows and changes. To promote the use of electric cars and build a more sustainable and effective transportation system from now on, manufacturers and other stakeholders are aggressively tackling these issues (Corradi et al. 2023).

Adherence to Regulations in Electric Car Technology

The Environment of Regulations

An essential component of electric vehicle (EV) technology is regulatory compliance, guaranteeing that EVs adhere to all applicable safety and environmental regulations. While EV regulations differ from nation to nation, there are a few universal areas of concern:

Property Guidelines

EVs must abide by safety requirements set by organizations and governments. These standards address handling high-voltage systems, occupant safety, and crashworthiness. Safety laws make electric vehicles (EVs) at least as safe as conventional cars.

Carbon Emissions and Fuel Economy

Vehicle emissions and fuel efficiency requirements are defined by regulatory organizations. Due to their reduced or zero exhaust emissions, EVs are frequently preferred. Regulations and incentives aimed at promoting environmentally friendly transportation stimulate the use of electric cars.

Efficiency of Energy

Vehicle manufacturers are required by law in some areas to adhere to specific energy efficiency criteria. This may encourage advancements in the efficiency and design of electric cars.

Safety of Batteries

Regulations cover battery safety, storage, and transportation because of the possible risks connected to lithium-ion batteries. These guidelines are intended to reduce battery manufacturing, use, and disposal hazards.

Impact on the Environment

It is frequently mandatory for EV manufacturers to evaluate and disclose the environmental effects of their automobiles. This includes life cycle evaluations considering how EVs affect the Environment at every stage, from manufacture to disposal.

The Function of State Grants

Numerous countries provide incentives to encourage the use of electric vehicles. Tax credits, refunds, lowered registration costs, and the availability of carpool lanes are a few examples of these incentives. The purpose of incentives is to increase consumer appeal and hasten the adoption of electric vehicles. Some governments are considering enacting Zero Emission Vehicle (ZEV) requirements, which compel automakers to sell a specific proportion of electric vehicles in addition to incentives. Due to these obligations, automakers are under regulatory pressure to invest in and develop electric cars.

International Harmonization

The global character of the electric car sector has prompted the worldwide harmonization of legislation and standards. Organizations such as the United Nations Economic Commission for Europe (UNECE) have attempted to produce worldwide technical norms for electric cars to standardize safety and environmental requirements across many locations. Harmonization makes it easier for manufacturers who build electric vehicles for the global market to comply.

Automobile Technology Testing and Validation

The Significance of Validation and Testing

In the development of electric vehicles (EVs), testing and validation are essential procedures. They guarantee that EVs adhere to performance, safety, and legal requirements. Various tests and evaluations covering different facets of the vehicle's construction and functioning are part of these procedures.

Safety Validation and Crash Testing

An essential part of confirming the safety of electric vehicles is crash testing. It entails modeling several accident situations to evaluate the vehicle's safety systems and construction performance. To safeguard occupants in the case of an accident, electric cars must adhere to safety regulations comparable to those for conventional vehicles. High-voltage electrical systems are also tested as part of safety validation to make sure they are built to endure mishaps without endangering passengers or first responders. These tests evaluate how the high-voltage system behaves both during and after a collision, as well as the structural integrity of the battery container.

Thermal Management and Battery Testing

Because the battery pack is the primary component of electric cars, battery testing and validation are essential. These tests evaluate the battery's functionality, dependability, and safety in various scenarios, such as heavy electrical loads and freezing temperatures. Testing must take battery temperature management into account. The process entails evaluating the efficiency of heating and cooling systems to guarantee that the battery functions within the designated temperature range. Effective thermal control reduces the possibility of thermal runaway and increases battery life.

Testing for Efficiency and Performance

Electric car economy and performance are also tested and validated. This entails assessing factors including handling, braking, acceleration, and energy usage. Manufacturers conduct these tests to maximize vehicle performance and guarantee that it meets or exceeds consumer expectations.

Testing for Regulatory Compliance

To make sure they meet pollution regulations and other legal criteria, electric cars have to go through a rigorous testing process. These tests verify that the

EV satisfies the relevant regulatory requirements by evaluating the vehicle's energy efficiency, safety features, pollution levels, and other characteristics.

Testing for Durability and Reliability

Testing for durability and dependability evaluates the long-term performance of electric cars. These tests replicate the deterioration that a vehicle could have during its lifespan. They entail putting the car through various stresses and settings, such as high mileage situations, temperature fluctuations, and vibrations from the road.

User Experience Testing

User experience testing assesses an electric vehicle's usability, comfort, and convenience from the viewpoints of the driver and passenger. This entails evaluating cabin comfort, infotainment systems, interior design, and user-friendliness. Automakers may improve their automobiles using user experience testing to satisfy customer demands and expectations better.

Upcoming Advances in Electric Vehicle Technology Technology

Advances in Battery Technology

Leading the way in the development of electric vehicles is still battery technology. Energy density, charging speed, and durability are improving due to battery chemistry, materials, and design developments. Positive advancements consist of:

- **Batteries with Solid State:** Compared to conventional lithium-ion batteries, solid-state batteries can provide more excellent safety, longer cycle life, quicker charging times, and better energy densities.
- **Batteries with Lithium-Sulfur:** The potential of lithium-sulfur batteries to improve energy density significantly and boost electric cars' range is the subject of ongoing study.

- **Sophisticated Cathode and Anode Materials:** Research is always conducted on cutting-edge materials like silicon and high-nickel cathodes to enhance battery performance.

Expansion of the Charging Infrastructure

The infrastructure for charging them must be expanded to increase the practicality and convenience of electric vehicles. Future advancements in the infrastructure for charging include:

- *Ultra-fast charging:* As ultra-fast charging networks are developed, electric car recharging will be possible quickly, considerably cutting down on charging periods.
- *Charging Wirelessly:* Wireless charging technology is anticipated to increase, providing electric car owners with simple, cable-free charging.

Connected and Autonomous Features

Electric vehicles are becoming more and more equipped with connected and autonomous functions, which improve user experience and safety. Advances in this field consist of:

- *Advanced Driver-Assistance Systems (ADAS):* As ADAS technology advances, cars will soon have fully autonomous driving thanks to features like automated parking, lane-keeping assistance, and adaptive cruise control.
- *Vehicle-to-Everything (V2X) Communication:* V2X technology makes it possible for cars to talk to infrastructure and each other, enhancing safety and traffic flow.

Vehicle Design and Lightweight Materials

Vehicle design is progressing due to developments in lightweight materials such as composites, aluminum, and carbon fiber. Electric cars may be made lighter, which improves performance and efficiency.

Recycling and Sustainability

Research and development on the sustainability of electric cars is a continuous process. Future advancements in this field of technology include:

- **Reusing Batteries:** Improved recycling technologies are being developed to recover valuable materials from end-of-life batteries and minimize waste and environmental damage.
- **Sustainable Materials:** To lessen the carbon footprint of producing electric vehicles, research on environmentally friendly and sustainable materials for vehicle manufacture is ongoing (Yisheng et al. 2023).

Electric Vehicle Technology User Experience

Comfort for Drivers and Passengers

Comfort for drivers and passengers is vital to the electric vehicle user experience. Electric vehicles frequently have quieter interiors because internal combustion engines produce no noise inside the car. A pleasant ride is also the goal of well-designed interiors and upgraded suspension systems.

Connectivity and Infotainment

Modern entertainment and communication systems are frequently seen in electric cars. The user experience is improved via voice-activated controls, smartphone connectivity, touchscreen displays, and over-the-air software upgrades. These amenities provide entertainment and convenience, improving the driving experience.

Mitigation of Anxiety

Resolving range anxiety is critical to user contentment. The fear of running out of power while traveling is lessened by the development of electric cars with greater ranges and improved route planning and range prediction algorithms.

Convenience of Charging

The ease of charging also affects user experience. This covers the accessibility of fast charging choices, charging stations, and payment alternatives. The ease of recharging an electric car has increased with the advent of sophisticated charging networks, which include wireless charging and ultra-fast chargers.

Remote Control and App Integration

Mobile apps that provide remote vehicle control, preconditioning, and monitoring are becoming increasingly popular. Through smartphone applications, users may improve convenience and user experience by checking battery status, controlling climatic settings, and locating local charging stations.

Driver Assistance and Autonomy

The user experience of electric vehicles is expanding to include technologies like automatic parking, lane-keeping assistance, and adaptive cruise control as they become more self-sufficient. These innovations improve safety while lightening the effort for drivers.

Eco-Driving and Environmental Awareness

Makers of electric vehicles frequently include eco-friendly driving features. To promote ecologically friendly driving practices, users may track how efficiently they drive and get advice on extending their electric range.

Regulation compliance guarantees that electric cars adhere to environmental and safety norms, and testing and validation are essential procedures to confirm performance, security, and regulatory compliance. Future innovations in electric car technology will encompass improvements in autonomous functions, lightweight materials, sustainability, charging infrastructure, and battery technology. The comfort of the driver and passengers, infotainment, range anxiety reduction, ease of charging, app integration, autonomous functions, and assistance for eco-driving are all included in the electric vehicle user experience. The continued focus on safety, performance, user experience, and sustainability in electric vehicle technology will significantly aid the adoption of electric cars and the development of the transportation landscape (Benaida, 2023).

Conclusion

In summary, batteries are the unsung heroes that power everything in our contemporary world, including the electric cars that promise a cleaner future and our smartphones. Even while lithium-ion batteries have fundamentally changed the electronics industry, much work must be done to improve battery technology, and new alternatives hold a lot of promise. Key factors influencing the development of energy storage include energy density, safety, conductive bridge electrolytes, battery management systems, and sustainability. Modern batteries are becoming increasingly popular in commercial and industrial settings, helping to create a more sustainable and clean future. Battery performance depends on cycle life, cost, and power density, particularly in electric cars. Optimizing efficiency and user-friendliness requires finding the ideal balance between weight, range, and charging speed. There is a lot of potential, including developments in battery chemistry, infrastructure construction, and electric car technology—all while upholding a strong commitment to regulatory compliance, safety, and scalability. Thanks to technological advancements, users may enjoy an experience with electric vehicles that is more sustainable and efficient in the future.

References

Ahasan Habib, A. K. M., Mohammad Kamrul Hasan, Ghassan F. Issa, Dalbir Singh, Shahnewaz Islam, Taher M. Ghazal, *Lithium-Ion Battery Management System for Electric Vehicles: Constraints, Challenges, and Recommendations, Batteries* 2023, 9(3), 152.

Chiara Corradi, Edgardo Sica, Piergiuseppe Morone, What drives electric vehicle adoption? Insights from a systematic review on European transport actors and behaviours, *Energy Research & Social Science*, 95, 2023, 102908.

Fuwei Xiang, Fang Cheng, Yongjiang Sun, Xiaoping Yang, Wen Lu, Rose Amal & Liming Dai, Recent advances in flexible batteries: From materials to applications, *Nano Research* 16, 2023, 4821–4854.

Jinqiu Zhou, Siyi Qian, Baojiu Hao, Jie Liu, Xi Zhou, Chenglin Yan, Tao Qian, Small Molecules, Great Powers: Chemistry of Small Organo-Chalcogenide Molecules in Rechargeable Li-Sulfur Batteries, *Advanced Functional Materials*, 33(25), 2023, 2213966.

Junfei Zhang, Yunling Wu, Miao Liu, Lu Huang, Yanguang Li, Yingpeng Wu, Self-Adaptive Re-Organization Enables Polythiophene as an Extraordinary Cathode Material for Aluminum-Ion Batteries with a Cycle Life of 100 000 Cycles, *Angewandte Chemie*, 135(8), 2023, e202215408.

Mohamed Benaida, Developing and extending usability heuristics evaluation for user interface design via AHP, *Soft Computing* 27, 9693–9707 (2023).

Un-Hyuck Kim, Soo-Been Lee, Ji-Hyun Ryu, Chong Seung Yoon, Yang-Kook Sun, Optimization of Ni-rich Li[Ni0.92−xCo0.04Mn0.04Alx]O2 cathodes for high energy density lithium-ion batteries, *Journal of Power Sources*, 564, 2023, 232850.

Yao Lv, Shifei Huang, Sirong Lu, Tianqi Jia, Yanru Liu, Wenbo Ding, Xiaoliang Yu, Feiyu Kang, Jiujun Zhang, Yidan Cao, Engineering of cobalt-free Ni-rich cathode material by dual-element modification to enable 4.5 V-class high-energy-density lithium-ion batteries, *Chemical Engineering Journal*, 455, 2023, 140652.

Yisheng An, Yuxin Gao, Naiqi Wu, Jiawei Zhu, Hongzhang Li, Jinhui Yang, Optimal scheduling of electric vehicle charging operations considering real-time traffic condition and travel distance, *Expert Systems with Applications*, 213, Part B, 2023, 118941.

Yongsheng Shi, Peipei Yin, Jun Li, Xiaozhuo Xu, Qinting Jiang, Jiayin Li, Hirbod Maleki Kheimeh Sari, Jingjing Wang, Wenbin Li, Junhua Hu, Qingxin Lin, Jingqian Liu, Jun Yang, Xifei Li, Ultra-high rate capability of in-situ anchoring FeF3 cathode onto double-enhanced conductive Fe/graphitic carbon for high energy density lithium-ion batteries, *Nano Energy*, 108, 2023, 108181.

Zhifan Zhang, Hailong Li, Longkan Wang, Guiyong Zhang, Zhi Zong, Shenhe Zhang, Protection mechanism of underwater double-hull coated with UHMW-PE subjected to shaped charge, *Ocean Engineering*, 272, 2023, 113842.

About the Editors

Dr. Suresh Nagappan Sundaram received his B.E. degree in Electrical and Electronic Engineering and M.E. degree in Control and Instrumentation in the Department of Electrical and Electronic Engineering from Anna University in the year 2007 and 2011 respectively. He did his doctorate degree in Full Time mode from National Institute of Technology, Tiruchirappally in 2019. He worked for two years in industry at various roles as designer, installation and service technician etc. He served as Assistant Professor in engineering colleges and presently working as Associate Professor in Electrical and Electronics Engineering Department at Saveetha School of Engineering, Chennai. His research interest includes microgrid, smart-grid, renewable energy, etc.

Mrs. N. S. Padmavathy, is currently pursuing her PhD at National Institute of Technology, Tiruchirappalli, India. She completed her Bachelor of Technology in Civil Engineering from VLB Janakiammal College of Engineering and Technology, Coimbatore, India in 2008 and Master of Technology in Structural Engineering from Kumaraguru College of Technology, Coimbatore, India in 2010. She has worked as Assistant Professor in Civil Engineering Department in Kumaraguru College of Technology, Coimbatore from 2010 to 2012. She is a life member in Indian Geotechnical Society. She has been a project student at ISRO Trivandrum, Kerala, India for 4 months in 2009-2010. Her Research area includes Structural designing, Composite materials, Concrete, Seismic Analysis, Transportation, Software analysis and design of Structures. In addition, she has published more than 5 articles in high quality international peer reviewed journals, 2+ book Chapters and also presented research papers at national/international conferences. Till now she has received one patent.

Dr. Mohit Hemath Kumar completed his B.E. (Mechanical Engineering) from Anna University, Chennai, Tamilnadu, India in the year 2012, and his

M.E. (Thermal Engineering with specialization in Refrigeration and Air Conditioning Engineering) from College of Engineering Guindy Campus, Anna University, Chennai in the year 2014. He completed his Ph.D. from Department of Mechanical Engineering, National Institute of Technology Tiruchirappalli in the year 2019 and he finished his Post-Doctoral Research from King Mongkut's University of Technology North Bangkok, Thailand. Currently, he is working as an Assistant Professor at Alliance University Bengaluru. He is a life member of the Institution of Engineers (India), the Powder Metallurgy Association of India (India), the Society for Failure Analysis (India), and the Science and Technology Research Association (Singapore). In addition, he has published more than 20 articles in high-quality international peer-reviewed journals, 20+ book Chapters, three books as Editor, and also presented research papers at national/international conferences. Till now he has received 14 patents and 18 industrial designs from Indian Intellectual Property Rights. His current research areas include Natural Fibre and Inorganic fillers Based Polymer Nanocomposites, Metal Matrix Composites, and Waste to Innovative Industrial Products Design.

Dr. A. Joseph Godfrey received his B.E. degree in Electrical and Electronics from Sun College of Engineering and Technology, Erachakulam, affiliated with Anna University, Chennai, India, in 2007 and M.Tech. Degree in Power Electronics and Drives from VIT University, Vellore, India, in 2009. From 2009 to 2011, he was a Lecturer at Narayanaguru College of Engineering, Manjaalumoodu, India. He completed his Ph.D. in Electrical and Electronics Engineering from the National Institute of Technology, Tiruchirappalli, India, in 2023. He works as an Assistant Professor at St. Joseph Engineering College, Mangalore, India. His research interests include the application of power electronics in electric vehicles and regenerative braking.

Index

S

T

U

V

W